M336
Mathematics and Computing: a third-level course

GROUPS & GEOMETRY

UNIT IB2
GROUPS: PROPERTIES AND EXAMPLES

Prepared for the course team by
Bob Margolis

The Open University

This text forms part of an Open University third-level course.
The main printed materials for this course are as follows.

Block 1
Unit IB1 Tilings
Unit IB2 Groups: properties and examples
Unit IB3 Frieze patterns
Unit IB4 Groups: axioms and their consequences

Block 2
Unit GR1 Properties of the integers
Unit GR2 Abelian and cyclic groups
Unit GE1 Counting with groups
Unit GE2 Periodic and transitive tilings

Block 3
Unit GR3 Decomposition of Abelian groups
Unit GR4 Finite groups 1
Unit GE3 Two-dimensional lattices
Unit GE4 Wallpaper patterns

Block 4
Unit GR5 Sylow's theorems
Unit GR6 Finite groups 2
Unit GE5 Groups and solids in three dimensions
Unit GE6 Three-dimensional lattices and polyhedra

The course was produced by the following team:

Andrew Adamyk (BBC Producer)
David Asche (Author, Software and Video)
Jenny Chalmers (Publishing Editor)
Bob Coates (Author)
Sarah Crompton (Graphic Designer)
David Crowe (Author and Video)
Margaret Crowe (Course Manager)
Alison George (Graphic Artist)
Derek Goldrei (Groups Exercises and Assessment)
Fred Holroyd (Chair, Author, Video and Academic Editor)
Jack Koumi (BBC Producer)
Tim Lister (Geometry Exercises and Assessment)
Roger Lowry (Publishing Editor)
Bob Margolis (Author)
Roy Nelson (Author and Video)
Joe Rooney (Author and Video)
Peter Strain-Clark (Author and Video)
Pip Surgey (BBC Producer)

With valuable assistance from:

Maths Faculty Course Materials Production Unit
Christine Bestavachvili (Video Presenter)
Ian Brodie (Reader)
Andrew Brown (Reader)
Judith Daniels (Video Presenter)
Kathleen Gilmartin (Video Presenter)
Liz Scott (Reader)
Heidi Wilson (Reader)
Robin Wilson (Reader)

The external assessor was:
Norman Biggs (Professor of Mathematics, LSE)

The Open University, Walton Hall, Milton Keynes, MK7 6AA.

First published 1994. Reprinted 1998, 2001 , 2005.

Copyright © 1994 The Open University

All rights reserved. No part of this publication may be reproduced, stored in a retrieval system or transmitted in any form or by any means, without written permission from the publisher or a licence from the Copyright Licensing Agency Limited. Details of such licences (for reprographic reproduction) may be obtained from the Copyright Licensing Agency Ltd of 90 Tottenham Court Road, London, W1P 9HE.

Edited, designed and typeset by the Open University using the Open University TeX System.

Printed in Malta by Gutenberg Press Limited.

ISBN 0 7492 2160 7

This text forms part of an Open University Third Level Course. If you would like a copy of *Studying with The Open University*, please write to the Central Enquiry Service, PO Box 200, The Open University, Walton Hall, Milton Keynes, MK7 6YZ. If you have not already enrolled on the Course and would like to buy this or other Open University material, please write to Open University Educational Enterprises Ltd, 12 Cofferidge Close, Stony Stratford, Milton Keynes, MK11 1BY, United Kingdom.

CONTENTS

Study guide	4
Introduction	5
1 Axioms and examples	**6**
1.1 The group axioms	6
1.2 Examples of groups	8
2 Subgroups	**16**
2.1 The subgroup axioms	16
2.2 Examples of subgroups	18
3 Generating subgroups	**21**
4 Cyclic groups	**24**
4.1 Definition and examples	24
4.2 Subgroups of cyclic groups	26
5 Group actions	**27**
Solutions to the exercises	33
Objectives	51
Index	52

STUDY GUIDE

In this unit, we assume that you have some familiarity with the idea of groups from your previous studies. With this in mind, in some places in this unit the discussion is fairly informal, the more formal development being postponed to *Unit IB4* (and in some cases to later in the Groups stream of the course).

One of the aims of this unit is to introduce you to several examples of groups that will recur later in the course. It is particularly important, therefore, that you take the time to work through the exercises in the unit, as this is the best way of becoming thoroughly familiar with these examples of groups.

The sections of this unit are not of equal length. Because of the number of examples of groups that it introduces, Section 1 comprises rather more than one-fifth of the work of this unit. However, to balance this, Sections 2–5 are rather shorter.

There is no video or audio programme associated with this unit.

You will not need the *Geometry Envelope* in your study of this unit.

INTRODUCTION

The purpose of this unit is to review the definition of a group, to review the basic properties of groups and to consider a number of examples of groups. The main content is the review of basic concepts connected with groups. However, the examples are important too, since they are typical of the groups that you will meet in the rest of the course.

Some of the groups that we discuss are groups of symmetries of various geometric objects, including some rather simple tilings and friezes. These are introduced because it is these groups of symmetries that will be the tools that enable us later on to classify friezes, wallpaper patterns and certain types of tilings.

Friezes are defined formally in Unit IB3, and *wallpaper patterns* in Unit GE4. You met *tilings* in Unit IB1.

We shall also look at some examples of cyclic and other Abelian groups. Cyclic groups form the 'atomic' building blocks for the Abelian groups, whose classification forms another major theme of the course.

The course views groups from two slightly different angles, corresponding roughly to geometric and algebraic viewpoints. For example, we may view the *Klein group*, V, as being defined by its Cayley table:

The notation V comes from the German *Vierergruppe* ('four-group').

$\circ$	e	a	b	c
e	e	a	b	c
a	a	e	c	b
b	b	c	e	a
c	c	b	a	e

On the other hand, the symmetries of a rectangle, $\Gamma(\square)$, form a group which is essentially the same as the Klein group. Which notation we use will tend to depend on whether the context is geometric or algebraic.

There is a further practical problem: the notation

Γ(picture of geometric object)

is very evocative but can be space consuming if the object is large. We shall, therefore, frequently introduce more compact and convenient notation for symmetry groups.

In Section 1 we remind you of the *axioms for a group* and look at some *examples of groups*. We shall also consider how to describe large finite and infinite groups, given that we cannot use a Cayley table as is sometimes done for small finite groups. This need to describe large finite and infinite groups leads to the concept of a set of *generators* for a group.

Section 2 recalls the definition of *subgroups* and considers some *examples* from the groups introduced in Section 1. The main aim of the section is to increase your familiarity with the example groups and their subgroups. The approach is largely *geometric*.

In Section 3, we take a second look at the *subgroups* from Section 2, this time taking a more *algebraic* approach to the description of the subgroups.

Cyclic groups form the main building blocks used in the Groups stream of the course. In Section 4 we remind you about *cyclic groups* and discuss some of their properties.

Finally, in Section 5, we develop the concept of *group actions*. In particular, we look at some examples that will be particularly useful in establishing results about groups later on in the course.

One central idea in this unit is that of generating a group or subgroup by specifying both a set of elements of the group (or subgroup) and relations between the elements of the generating set. This idea is introduced informally, through specific examples. However, formalizing the idea of generators for a group, and ensuring that the calculations that 'obviously' lead to a group or subgroup are valid in general, is a task that requires some care. As we remarked in the Study Guide, we postpone this formalization until *Unit IB4*.

1 AXIOMS AND EXAMPLES

In this section we remind you of the axioms for a group and look at some examples appropriate for the rest of the course. Some of the examples may be familiar to you, some may not be. Since the course demands real familiarity with many of these examples, we ask you to work carefully through *all* the exercises.

1.1 The group axioms

As you may remember from your previous studies, the group axioms are an abstraction of the common properties of various sets of numbers, functions, transformations, etc., together with various operations for combining the elements of these sets. This abstraction process leads to the following definition.

Definition 1.1 Group axioms

A set G together with a binary operation $\circ$ defined on G is a **group** if the following axioms are satisfied.

Closure For all pairs x, y of elements in G, $x \circ y$ is also in G.

Associativity For all choices of three elements x, y, z in G, we have
$$(x \circ y) \circ z = x \circ (y \circ z).$$

Identity There is an **identity element** e, say, in G with the property that, for any element x in G,
$$e \circ x = x = x \circ e.$$

Inverses For each element x in G, there is an **inverse element** x^{-1} in G with the property that
$$x \circ x^{-1} = e = x^{-1} \circ x,$$
where e is the identity element of G.

A couple of comments are in order on the way the axioms have been stated above.

Firstly, we should really say that $(G, \circ)$ is a group if the axioms are satisfied; that is, we should always make explicit reference to the set *and* the operation when talking about groups. However, such precision leads to clumsiness, and we shall follow convention by referring only to the set whenever there is a clearly understood operation that makes the set into a group. We shall also usually write the group operation as if it were multiplication, writing xy where we should, strictly, write $x \circ y$. The main exception to this rule will be where the group operation is addition.

Secondly, you may have noticed that the phrase 'an identity' in the third axiom has become 'the identity' in the fourth. This implies that a group can only have one identity, a fact that we shall establish formally in *Unit IB4*.

In the following exercises we ask you to check that each of the given sets, with the given operation, forms a group. You may make reasonable assumptions about properties of familiar operations on numbers. For each example, you should specify the identity of the group and how to find an inverse for each element.

Exercise 1.1

Set: $\mathbb{Z}$ (the integers).
Operation: $+$ (addition).

Exercise 1.2

Set: all 2×2 matrices with real numbers as entries and non-zero determinant.
Operation: matrix multiplication.
(You may assume that matrix multiplication is associative.)

This group is usually called $GL(2, \mathbb{R})$. The GL is for *general linear*, referring to the fact that matrices represent linear transformations between vector spaces. The 2 refers to the size of the matrices and the $\mathbb{R}$ to the fact that the entries are real numbers.

Exercise 1.3

Set: all 2×2 matrices of the form
$$\begin{bmatrix} 1 & a \\ 0 & 1 \end{bmatrix}, \quad a \in \mathbb{Z}.$$

Operation: matrix multiplication.
(You may assume that matrix multiplication is associative.)

The above exercises provide three examples of groups that are of importance in this course. A further important example, of a whole class of groups, is provided by the groups S_n (where n is a positive integer). The formal definition of S_n is the set of all *one–one* functions from the set

$$\{1, \ldots, n\}$$

onto itself, with composition of functions as the operation. The elements of each S_n are called *permutations*, and each S_n is an example of a *permutation group*.

We remind you of the notations used for permutations by giving a specific example. In S_4, the *two-line notation*

$$\begin{pmatrix} 1 & 2 & 3 & 4 \\ 4 & 3 & 2 & 1 \end{pmatrix}$$

represents the following permutation:

$1 \mapsto 4$
$2 \mapsto 3$
$3 \mapsto 2$
$4 \mapsto 1$

Since, for this permutation, $1 \mapsto 4 \mapsto 1$ and $2 \mapsto 3 \mapsto 2$, it can be written in *cycle notation* as

$$(1\,4)(2\,3).$$

Notice that the cycles $(1\,4)$ and $(2\,3)$ are *disjoint* (i.e. they contain no numbers in common). A permutation written in terms of disjoint cycles is said to be in **cycle form**.

Exercise 1.4

Evaluate the following product in S_5:

$$\begin{pmatrix} 1 & 2 & 3 & 4 & 5 \\ 3 & 5 & 4 & 1 & 2 \end{pmatrix} \begin{pmatrix} 1 & 2 & 3 & 4 & 5 \\ 3 & 5 & 1 & 2 & 4 \end{pmatrix}$$

Leave your answer in two-line notation.

Exercise 1.5

Write the following element of S_5 in cycle form:

$$\begin{pmatrix} 1 & 2 & 3 & 4 & 5 \\ 3 & 5 & 4 & 1 & 2 \end{pmatrix}$$

Exercise 1.6

Write the following product in S_6 in cycle form:

$$(1\ 6\ 3)(2\ 5\ 4)(1\ 3\ 5\ 6\ 2)$$

To close this subsection, we want to say two things about associativity.

Firstly, the associativity axiom guarantees that the two ways of bracketing a combination of *three* elements leads to the same result. It says nothing directly about expressions such as

$$x_1 \circ x_2 \circ \cdots \circ x_n.$$

Fortunately, it is possible to prove (by the Principle of Mathematical Induction) that all ways of bracketing such an expression lead to the same result. The proof is more involved than illuminating, and so we do not give it. We shall assume the result and write such expressions without brackets, doing the calculation by combining pairs in the most convenient way.

Secondly, a 'first principles' proof that associativity holds in a particular example is usually the messiest part of verifying that an example is a group. Fortunately, for very large classes of groups, including most of those considered in this course, we can deal with associativity once and for all as follows. For many examples of groups, like the groups of matrices above, the elements of the set are (or can be treated as) *functions*. For example, the 2×2 matrices in $\text{GL}(2, \mathbb{R})$ can be thought of as being functions from the plane $\mathbb{R}^2$ to itself. The group operation in such cases is, effectively, composition of functions. Now, as you may remember from your previous studies, composition of functions is always associative. This fact removes the need to check associativity in a very large number of cases. We shall use this principle freely in the rest of the course.

1.2 Examples of groups

We now want to consider some examples of groups, both finite and infinite, which are typical of the groups encountered in the Geometry stream of the course. The examples also serve to introduce our main method of describing groups.

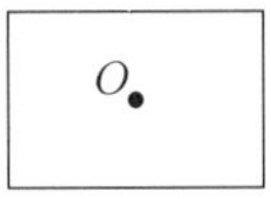

The first example is the *group of symmetries of the rectangle* shown in Figure 1.1. This group consists of the isometries of the plane ($\mathbb{R}^2$) that leave the rectangle 'fixed'. That is, it is the set of isometries that map the set of points forming the rectangle onto itself. Note that we have *not* said that the isometries fix each point of the rectangle. For example, the isometry given by a half-turn (i.e. a turn through π) about the centre O of the rectangle leaves no point of the rectangle in its original position, but it does fix the whole rectangle and so is a symmetry of the rectangle.

Figure 1.1

Isometries were defined in *Unit IB1*.

We regard a rectangle as consisting of just its boundary.

Although we are sure that you are familiar with the symmetries of the rectangle, it will be useful to go into a little detail as to how we can be sure exactly which isometries belong to the group. To do so, we shall use an approach that will be useful in more complicated examples; the approach is based on consequences of the defining property of isometries: that they preserve distance. We find all the symmetries of the rectangle in two stages. Firstly, we decide what isometries might possibly occur as symmetries. Secondly, we show that all such possibilities can be realized.

The strategy adopted here is a very useful one. We find an upper limit to the number of isometries in a symmetry group and then show that we can find that many different symmetries.

If we think about the upper long side of the rectangle, under an isometry it can only be mapped to itself or to the bottom long side. (That is because the ends of the upper long side must be the same distance apart after mapping as before and because isometries always map straight lines to straight lines.) Similarly, the left short side can only be mapped to itself or to the right short side. Since a corner is the intersection of a long side and a short side, corners must map to corners. The top left corner has at most four possible images. Furthermore, once we decide which corner it maps to, this determines the images of the adjacent corners. So our decision as to where the top left corner maps determines the images of three non-collinear points and hence, by the Fundamental Theorem of Affine Geometry, the isometry. There are, therefore, *at most* four symmetries.

See Theorem 4.1 of Unit IB1.

However, we can actually find four different symmetries of the rectangle. The identity mapping is one; a half-turn about the point O is another; reflection in the horizontal axis of symmetry gives a third; reflection in the vertical axis of symmetry gives the last. Since we know that there are at most four, we know that we have them all.

If we denote the symmetries (in the order described above) as e, r, h and v, then we can construct a Cayley table quite easily:

∘	e	r	h	v
e	e	r	h	v
r	r	e	v	h
h	h	v	e	r
v	v	h	r	e

By choosing an origin at the centre of the rectangle and coordinate axes parallel to the sides of the rectangle, we can represent the four symmetries e, r, h and v by the matrices

$$\begin{bmatrix} 1 & 0 \\ 0 & 1 \end{bmatrix}, \begin{bmatrix} -1 & 0 \\ 0 & -1 \end{bmatrix}, \begin{bmatrix} 1 & 0 \\ 0 & -1 \end{bmatrix} \text{ and } \begin{bmatrix} -1 & 0 \\ 0 & 1 \end{bmatrix}$$

respectively. Thus we have a second representation of the group of symmetries of the rectangle, this time as a group of matrices. Note that all four of these matrices are orthogonal (in the sense defined in *Unit IB1*), as we would expect.

See Theorem 3.2 of Unit IB1.

There is a third representation of this group, which can be obtained if we number the corners of the rectangle 1 to 4, as in Figure 1.2. Since, as we have observed, corners map to corners, each symmetry corresponds to a permutation of

$$\{1, 2, 3, 4\},$$

that is to an element of S_4, as follows:

$$e \text{ corresponds to } \begin{pmatrix} 1 & 2 & 3 & 4 \\ 1 & 2 & 3 & 4 \end{pmatrix}$$

$$r \text{ corresponds to } \begin{pmatrix} 1 & 2 & 3 & 4 \\ 3 & 4 & 1 & 2 \end{pmatrix} = (1\,3)(2\,4)$$

$$h \text{ corresponds to } \begin{pmatrix} 1 & 2 & 3 & 4 \\ 2 & 1 & 4 & 3 \end{pmatrix} = (1\,2)(3\,4)$$

$$v \text{ corresponds to } \begin{pmatrix} 1 & 2 & 3 & 4 \\ 4 & 3 & 2 & 1 \end{pmatrix} = (1\,4)(2\,3)$$

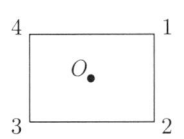

Figure 1.2

This group of 4 permutations forms only a subgroup of S_4, which has 24 elements.

These three representations illustrate a point that we made in the
Introduction: a group may have several different but equivalent
representations. Here, then, we can consider that we have three concrete
examples of groups all essentially the same as the Klein group, V. Thus,
depending on the context, we may regard any one of these as *the* Klein
group.

The above is an example of a whole class of symmetry groups. If we have a
subset X of the plane, $\mathbb{R}^2$, then the isometries of $\mathbb{R}^2$ that map X onto itself
form a group called the **symmetry group** of X. We will denote the
symmetry group of X by $\Gamma(X)$. Thus we denote the **symmetry group of
the rectangle** by

$\Gamma(\square)$.

Or, more generally still, $\mathbb{R}^n$.

An alternative notation is $S(X)$, but we shall use $\Gamma(X)$.

Where the subset X is large, or complicated, the $\Gamma(\ldots)$ notation can become
unwieldy. In such cases we shall introduce more convenient notation.

The second example is closely related to the first. It is the group, which we
denote by E_1, of symmetries of the pattern illustrated in Figure 1.3, which is
a rather elementary *frieze*, obtained by forming an infinite strip of the
rectangles like the one discussed above.

This is a case where the $\Gamma(\ldots)$ notation is too cumbersome.

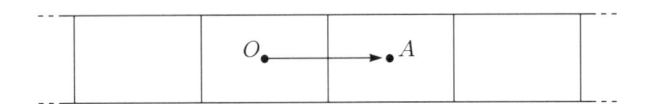

Figure 1.3

The situation is rather more complicated here, because there is an infinite
number of translations which preserve the infinite pattern. For example, the
translation shown by the arrow OA may be repeated as often as you like.
There is also an infinite number of rotations: a half-turn about the centre of
any short side or about the centre of any rectangle will map the frieze onto
itself.

In order to catalogue the isometries that make up E_1, the **symmetry
group of a plain rectangular frieze**, we use the same approach as in the
previous example. We first determine all the possible symmetries of the
frieze and then show that all these can be realized. We shall refer to the
rectangle centred at O as the *base* rectangle of the frieze.

Exercise 1.7

Describe the set of all possible images of the point O under a symmetry of
the frieze.

Exercise 1.8

Describe the set of all possible images of the top left corner of the base
rectangle under a symmetry of the frieze.

Exercise 1.9

Describe the set of all possible images of the top right corner of the base
rectangle under a symmetry of the frieze.

Once we know the images of the centre of the base rectangle and of both
ends of the top side, then the isometry is fixed. (This is because, by the
Fundamental Theorem of Affine Geometry, any isometry is fixed once you
know the images of three non-collinear points.) Therefore, the solutions to
the last three exercises enable us to describe the set of *possible* symmetries
of the frieze. We can now show that all these possibilities can actually be
achieved.

We begin by considering those isometries that fix O. All the isometries that make up $\Gamma(\square)$, when applied to the base rectangle, map not only that rectangle to itself, and hence O to itself, but also the whole frieze to itself. These four isometries correspond to the ends of the top side of the base rectangle mapping to themselves (either fixing both or interchanging them) and to both possibilities for mapping them to the ends of the bottom side of the base rectangle. There are no other isometries that fix O.

You may like to take a few minutes to convince yourself of this, possibly with the aid of a piece of tracing paper.

Now for the isometries that do not fix O. Any choice of images for the ends of the top side of the base rectangle can be achieved as follows. First map the ends of the top side of the base rectangle to the corners of the base rectangle that correspond to the wanted images, then apply a translation to move them to their final position at the corners of the required rectangle. For example, to map O, P and Q in Figure 1.4(a) to O', P' and Q', first apply v from $\Gamma(\square)$ to the base rectangle, to give the image shown in Figure 1.4(b), and then translate to the right by twice the length of the base rectangle to achieve the wanted result. All the isometries that do not fix O can be obtained in this way.

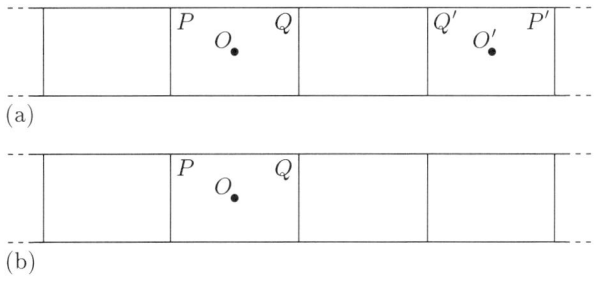

Figure 1.4

Therefore we can obtain all the elements of E_1 as an element of $\Gamma(\square)$ followed by a suitable translation. Since elements of $\Gamma(\square)$ are, effectively, orthogonal matrices, this description of the elements of E_1 corresponds to the standard form of isometries from *Unit IB1*.

Expressing the elements of $\Gamma(\square)$ as matrices, we have the standard form
$$t[\mathbf{p}]\lambda[\mathbf{A}]$$
as in *Unit IB1*.

We can, however, obtain a different description of E_1 in terms of rotations, reflections, translations and glide reflections.

We can map O to the centre of any chosen rectangle (and the ends of the top side of the base rectangle to the corresponding ends of the top side of the image of the base rectangle) by translating left or right by a suitable multiple of the length of a rectangle.

We can map O to the centre of any rectangle and reverse the sense of the top side by a reflection either in a short side of one of the rectangles, as shown in Figure 1.5(a), or in the vertical axis of symmetry of one of the rectangles, as shown in Figure 1.5(b).

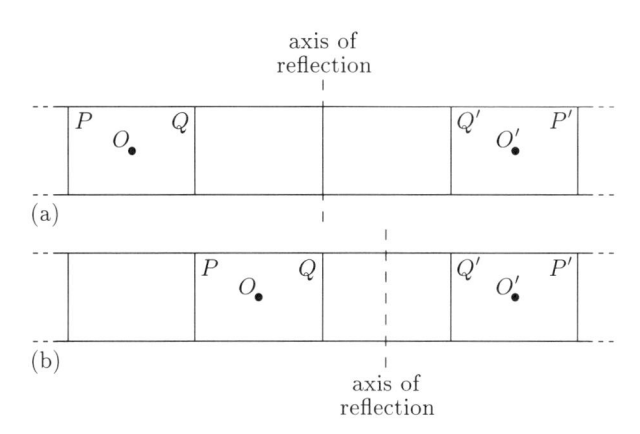

Figure 1.5

We can map O to any centre and the top side to the corresponding bottom side (but reversing the orientation) by a glide reflection in the horizontal axis of symmetry, as shown in Figure 1.6.

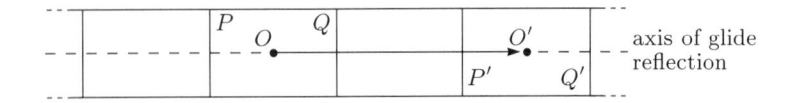

Figure 1.6

Finally, the symmetry obtained as in the last case, but preserving the orientation of the side, can be achieved by a half-turn either about the centre of a rectangle, as shown in Figure 1.7(a), or about the centre of a short side, as shown in Figure 1.7(b).

(a)

(b)

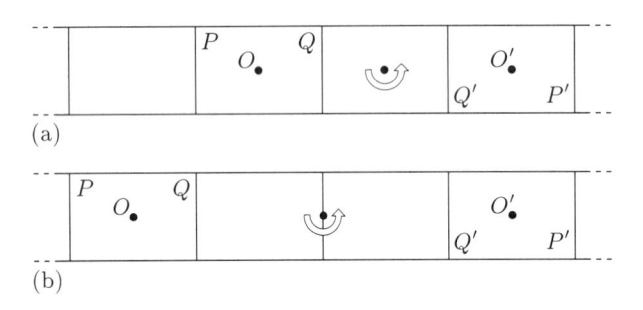

Figure 1.7

The above four cases encompass *all* the elements of E_1. Hence we have a description of E_1 in terms of translations, rotations, reflections and glide reflections.

Unlike the group $\Gamma(\square)$, it is not possible to provide a Cayley table for the group E_1 since it contains infinitely many elements. Instead, what we can do is to give a subset of E_1 and show that all the elements of the group can be expressed in terms of the elements of this subset. The clue as to how to do this is in our first, 'standard form' description of the elements of E_1, that is as elements of the form

an element of $\Gamma(\square)$ followed by a translation.

The element of $\Gamma(\square)$ must be e, r, h or v. The translation must be left or right by a multiple of the length of the base rectangle. As in *Unit IB1*, we denote a translation to the right by the vector **a** (shown in Figure 1.8) by $t[\mathbf{a}]$.

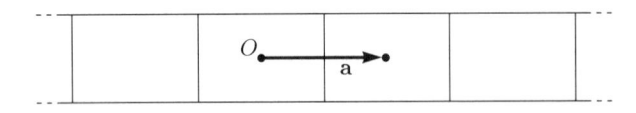

Figure 1.8

Any translation to the right can be achieved by taking enough copies of $t[\mathbf{a}]$, that is by $(t[\mathbf{a}])^n$, for some positive integer n.

Any translation to the left is the inverse of one to the right, and so can be achieved by $(t[-\mathbf{a}])^n = (t[\mathbf{a}])^{-n}$, for some positive integer n.

Thus all the translations may be described as

$(t[\mathbf{a}])^n$, for some *integer* n.

As usual, any element raised to the power zero is taken to be the identity element.

We can summarize this by saying

$$E_1 = \{(t[\mathbf{a}])^n \, x : n \in \mathbb{Z}, \, x \in \Gamma(\square)\}.$$

What we have achieved is to be able to express any element of the infinite group E_1 in terms of just five elements:

e, r, h, v and $t[\mathbf{a}]$.

We say that the set

$\{e, r, h, v, t[\mathbf{a}]\}$

generates E_1.

The notion of *generators* will be formalized in *Unit IB4*.

Expressing E_1 in terms of a set of generators is more than a way of *describing* the group. It also leads to a practical method of carrying out calculations with the group elements.

Exercise 1.10

By considering what happens to the base rectangle of a plain rectangular frieze, show that

(a) $r \, t[\mathbf{a}] = (t[\mathbf{a}])^{-1} \, r$;

(b) $h \, t[\mathbf{a}] = t[\mathbf{a}] \, h$;

(c) $v \, t[\mathbf{a}] = (t[\mathbf{a}])^{-1} \, v$.

Given the relations between the generators found in the solutions to Exercise 1.10, together with the Cayley table for $\Gamma(\square)$, we can express the composite of *any* two elements of E_1 in the 'standard form'

$$(t[\mathbf{a}])^n \, x, \quad n \in \mathbb{Z}, x \in \Gamma(\square),$$

that is with the translation part on the left.

For example,

$$\begin{aligned}(t[\mathbf{a}])^3 \, v \, (t[\mathbf{a}])^2 \, h &= (t[\mathbf{a}])^3 \, v \, t[\mathbf{a}] \, t[\mathbf{a}] \, h \\ &= (t[\mathbf{a}])^3 \, (v \, t[\mathbf{a}]) \, t[\mathbf{a}] \, h \\ &= (t[\mathbf{a}])^3 \, ((t[\mathbf{a}])^{-1} \, v) \, t[\mathbf{a}] \, h \\ &= (t[\mathbf{a}])^2 \, (v \, t[\mathbf{a}]) \, h \\ &= (t[\mathbf{a}])^2 \, ((t[\mathbf{a}])^{-1} \, v) \, h \\ &= t[\mathbf{a}] \, (v \, h) \\ &= t[\mathbf{a}] \, r.\end{aligned}$$

We have laboured the calculation somewhat in order to show how the relations between the generators have been used to simplify the composite.

Exercise 1.11

Use the relations between the generators to calculate the following composites of the elements $(t[\mathbf{a}])^2 \, r$ and $(t[\mathbf{a}])^3 \, v$ of E_1, expressing the answers in 'standard form':

(a) $(t[\mathbf{a}])^2 \, r \, (t[\mathbf{a}]^3) \, v$;

(b) $(t[\mathbf{a}])^3 \, v \, (t[\mathbf{a}])^2 \, r$.

We now note that the relations $r = hv = vh$ and $h^2 = v^2 = e$ are sufficient to draw up the Cayley table for $\Gamma(\square)$ and that

$$r \, t[\mathbf{a}] = v \, h \, t[\mathbf{a}] = v \, t[\mathbf{a}] \, h = (t[\mathbf{a}])^{-1} \, v \, h = (t[\mathbf{a}])^{-1} \, r,$$

so that the relation $r \, t[\mathbf{a}] = (t[\mathbf{a}])^{-1} \, r$ can be derived from the relations $r = vh$, $h \, t[\mathbf{a}] = t[\mathbf{a}] \, h$ and $v \, t[\mathbf{a}] = (t[\mathbf{a}])^{-1} \, v$. Therefore, if we add the relations $h \, t[\mathbf{a}] = t[\mathbf{a}] \, h$, $v \, t[\mathbf{a}] = (t[\mathbf{a}])^{-1} \, v$, $r = vh = hv$ and $h^2 = v^2 = e$ between the generators to the set description of E_1, we obtain a fairly compact statement of everything that we need in order to do calculations with the elements of E_1:

$$E_1 = \{(t[\mathbf{a}])^n\, x : n \in \mathbb{Z},\ x \in \Gamma(\square);$$
$$h\, t[\mathbf{a}] = t[\mathbf{a}]\, h,\ v\, t[\mathbf{a}] = (t[\mathbf{a}])^{-1}\, v,\ r = hv = vh,\ h^2 = v^2 = e\}.$$

It can be argued that the relations $r = hv = vh$ and $h^2 = v^2 = e$, being essentially a definition of $\Gamma(\square)$, are, really, redundant. We prefer to include them.

We finish this section with two more examples, one finite and one infinite. The finite one may be familiar to you.

The symmetries of a regular hexagon form a group, known as the **symmetry group of the regular hexagon**, which we shall refer to as D_6.

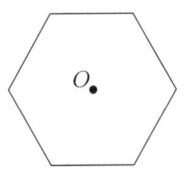

Figure 1.9 A regular hexagon with centre O.

You may have already met this group in your previous studies under the rather more suggestive name of $S(\hexagon)$. However, there are two other names in fairly common use. One is our preferred notation D_6, the suffix denoting the number of *sides* of the *polygon*; the other is D_{12}, the suffix indicating the number of *elements* in the *group*. The label D is for *dihedral*. There is a whole family of dihedral groups, one for each regular polygon.

Exercise 1.12

Explain why there are at most twelve symmetries of a regular hexagon and how all twelve can be realized.

A group with 12 elements, such as D_6, is really too large to deal with conveniently by a Cayley table. As with E_1 above, it is useful to look for a set of generators.

It is fairly easy to spot a possible generator for all the rotations: the rotation anticlockwise about the centre through $\pi/3$ can be used to produce all the rotations (including the identity).

Composition of rotations, with the same centre, always produces a rotation and never a reflection. As a consequence, a set of generators of D_6 must contain a reflection, since the whole group does. Consider, for example, the reflection in the line shown in Figure 1.10.

Actually, as you may wish to check, *any* reflection will do.

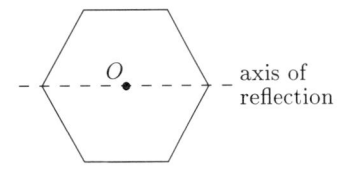

Figure 1.10

Let us label the chosen rotation r and the chosen reflection s. We shall now show that these two elements generate D_6.

In the notation of *Unit IB1*, $r = r[\pi/3]$ and $s = q[0]$. We have chosen shorter labels for convenience in the calculations which follow.

Exercise 1.13

Express each of the six rotations in D_6 as powers of r.

Exercise 1.14

Express each of the six reflections in D_6 in the form $r^n s$, for suitable integers n.

Exercise 1.15

Since sr must be a group element (by the closure axiom), it must be one of the elements found in the solutions to the previous two exercises. Which one?

From the solutions to the above exercises, we have a description of D_6 that will enable us to do calculations with its elements in the same way as with E_1 above. The group D_6 is generated by two elements r and s about which we know the following:

$$r^6 = s^2 = e, \quad sr = r^5 s.$$

The final relation can also be written $sr = r^{-1}s$.

We can give a complete and compact description of D_6, in the same way as we gave one for E_1, as follows:

$$D_6 = \{r^m s^n : m = 0, \ldots, 5, \ n = 0, 1; \ r^6 = s^2 = e, \ sr = r^5 s\}.$$

This description gives not only the 'standard form' of the elements,

$$r^m s^n, \quad m = 0, \ldots, 5, \ n = 0, 1,$$

Note that, since $r^0 = s^0 = e$, we usually write $r^m s^0$ as r^m and $r^0 s^n$ as s^n.

but also the relations,

$$r^6 = s^2 = e, \quad sr = r^5 s,$$

which are needed to reduce the product of two elements to this 'standard form'.

The final example is an extension of the group E_1 of symmetries of a plain rectangular frieze. We consider the group E_2 of symmetries of a plain rectangular tiling, part of which is shown in Figure 1.11.

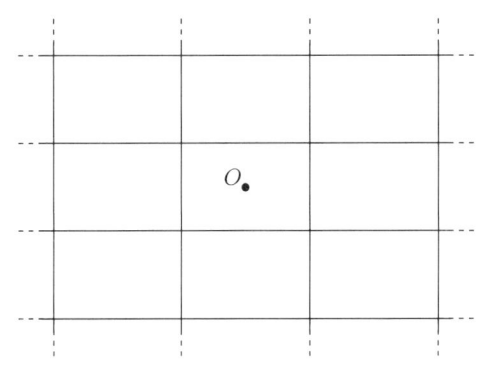

Figure 1.11

To obtain the elements of E_2, the **symmetry group of a plain rectangular tiling**, we can argue in much the same way as for the frieze. We consider what happens to the base rectangle, centred at O.

Firstly, the image of the base rectangle must be one of the other rectangles of the tiling, and furthermore there are four possible ways of orienting the image of the base rectangle in the required position. Secondly, all such images are achievable: a suitably chosen element of $\Gamma(\square)$ will get the orientation of the base rectangle correct, and a translation will then place it in the required image position. This time, however, the translation may be horizontal, vertical or a composite of both.

A set of generators for E_2 is, therefore,

$$\{e, r, h, v, t[\mathbf{a}], t[\mathbf{b}]\},$$

where $t[\mathbf{a}]$ and $t[\mathbf{b}]$ denote translation through the vectors $\mathbf{a}$ and $\mathbf{b}$ shown in Figure 1.12.

You may have noticed that since, in $\Gamma(\square)$, $r = hv$ and $h^2 = e$, for example, we need not have included all the elements of $\Gamma(\square)$ in the sets of generators for E_1 and E_2. In fact any two of r, h and v are sufficient. However, there is no harm in including 'redundant' generators.

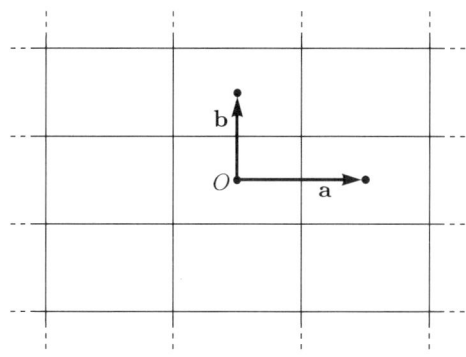

Figure 1.12

Exercise 1.16

Using a tracing of the plain rectangular tiling, or otherwise, verify the following relationships between the generators of E_2:

(a) $t[\mathbf{a}]\,t[\mathbf{b}] = t[\mathbf{b}]\,t[\mathbf{a}]$;
(b) $r\,t[\mathbf{a}] = (t[\mathbf{a}])^{-1}\,r$;
(c) $r\,t[\mathbf{b}] = (t[\mathbf{b}])^{-1}\,r$;
(d) $h\,t[\mathbf{a}] = t[\mathbf{a}]\,h$;
(e) $h\,t[\mathbf{b}] = (t[\mathbf{b}])^{-1}\,h$;
(f) $v\,t[\mathbf{a}] = (t[\mathbf{a}])^{-1}\,v$;
(g) $v\,t[\mathbf{b}] = t[\mathbf{b}]\,v$.

We now have a complete and compact description of E_2 as

$$E_2 = \{(t[\mathbf{a}])^m\,t([\mathbf{b}])^n\,x : m, n \in \mathbb{Z},\ x \in \Gamma(\square);$$
$$t[\mathbf{a}]\,t[\mathbf{b}] = t[\mathbf{b}]\,t[\mathbf{a}],\ r\,t[\mathbf{a}] = (t[\mathbf{a}])^{-1}\,r,\ r\,t[\mathbf{b}] = (t[\mathbf{b}])^{-1}\,r,$$
$$h\,t[\mathbf{a}] = t[\mathbf{a}]\,h,\ h\,t[\mathbf{b}] = (t[\mathbf{b}])^{-1}\,h,\ v\,t[\mathbf{a}] = (t[\mathbf{a}])^{-1}\,v,$$
$$v\,t[\mathbf{b}] = t[\mathbf{b}]\,v,\ r = hv = vh,\ h^2 = v^2 = e\}.$$

Note that we have included the 'redundant' relations that define $\Gamma(\square)$.

The four example groups in this subsection, together with the method of describing groups by specifying a set of generators and relationships between the generators, will recur throughout the course. In particular, two of the example groups will be generalized in the Geometry stream: E_1, which is an example of a *frieze group*, and E_2, which is an example of a *wallpaper group*. The idea of specifying a group by generators and relations is central to the Groups stream.

2 SUBGROUPS

In this section we review the ideas of subgroups and look at some examples obtained from the groups in Section 1.

2.1 The subgroup axioms

The underlying idea of a subgroup as a 'group within a group' is straightforward. For example, the group $\Gamma(\square)$ of symmetries of a rectangle appears within the group E_1 of symmetries of a plain rectangular frieze.

However, a precise definition requires just a little care. For a set H to be a subgroup of a group G, we require:

- H to be a subset of G, written $H \subseteq G$;
- H to be a group with respect to the *same* operation that makes G a group.

The first requirement is the 'sub' part of the idea. The second links the group structures in H and G. We may not define a new operation on H if it is to count as a subgroup, we must take the one that H inherits from G.

These requirements lead to the following formal definition.

Definition 2.1 *Subgroup axioms*

If G is a group with binary operation $\circ$ and H is a subset of G, then H is a **subgroup** of G (with binary operation $\circ$) provided that the following axioms are satisfied.

Closure For all pairs x, y of elements in H, $x \circ y$ is also in H.

Identity The identity element e of G is in H.

Inverses For each element x in H, the inverse element x^{-1} in G is also in H.

We include the binary operator $\circ$ here to emphasize that the binary operations of G and H must be the same. We shall in general, however, omit the operator and write xy rather than $x \circ y$.

The parts of this definition are similar to the corresponding parts of the definition of a group, but there are some subtle differences.

1 The closure axiom requires that H is closed *under the same operation as that in G*.

2 The identity axiom requires that the identity element (that we know G possesses) belongs to H.

3 The inverses axiom requires that the inverse of each element of H (which we know exists in G) belongs to H.

As you may have noted, the associativity axiom is 'missing'; this is the one group property that we *can* take as being inherited from G. Actually, this is a bit of an overstatement; what we can say is that once that we know that a subset is *closed* then the subset inherits associativity from the group as a whole. This is because if x, y and z are in the subset, we know that

$$(xy)z = x(yz),$$

by the associativity axiom for the group G, and we also know (by the closure axiom for subgroups) that both $(xy)z$ and $x(yz)$ are in the subset.

You should also note that, since the identity element of G must be in any subgroup, a subgroup cannot be empty.

Exercise 2.1

Find all the subgroups of the group $\Gamma(\square)$.

Hint There are not very many.

Exercise 2.2

This exercise concerns the group E_1 of symmetries of a plain rectangular frieze.

(a) Show that the set R of all rotations in E_1 does not form a subgroup of E_1.

(b) Show that the set T of all integer powers of $t[\mathbf{a}]$ is a subgroup of E_1.

Exercise 2.3

Show that, for any group G, the set $\{e\}$, consisting of just the identity element, and the set G itself are both subgroups of G.

We can deduce two useful results, from Exercises 2.2 and 2.3.

Lemma 2.1

(a) Any *non-trivial* group G always contains at least two subgroups: the *trivial* subgroup $\{e\}$, containing just the identity element, and the group G itself.

(b) For any group G and any element x in G, the set consisting of all integer powers of x is a subgroup of G.

The *trivial* group $G = \{e\}$ contains just one subgroup, namely $G = \{e\}$.

You may have previously seen such a subgroup described as the *cyclic* subgroup generated by the element.

We proved the first of these statements in the solution to Exercise 2.3. The proof of the second statement is identical to the proof of the special case $G = E_1, x = t[\mathbf{a}]$ given in the solution to Exercise 2.2. If you inspect that solution carefully, you will see that we did not make use of the particular nature of E_1 or of $t[\mathbf{a}]$.

2.2 Examples of subgroups

We now look at a number of other examples of subgroups, using our groups E_1, E_2 and D_6. Some of the methods that we use to find subgroups will generalize; these generalizations will be considered later in the course.

We begin by finding some subgroups of D_6, which we shall discuss using the description from Section 1:

$$D_6 = \{r^m s^n : m = 0, \ldots, 5,\ n = 0, 1;\ r^6 = s^2 = e,\ sr = r^5 s\}.$$

Remember, this gives the elements of D_6 as

$$e = r^0,\ r,\ r^2,\ r^3,\ r^4,\ r^5,\ s,\ rs,\ r^2 s,\ r^3 s,\ r^4 s,\ r^5 s.$$

One way of creating subgroups is to take an element of the group and calculate all its integer powers. By Lemma 2.1(b), this process will always give a subgroup, known as a *cyclic* subgroup. For example, taking r and calculating its integer powers gives the set

$$\{e = r^0,\ r,\ r^2,\ r^3,\ r^4,\ r^5\}.$$

Cyclic groups will be formally defined in Section 4.

No more different elements can be generated in this way because we know that $r^6 = e$.

The usual notation for subgroups generated in this way uses angle brackets:

$$\langle r \rangle = \{e = r^0,\ r,\ r^2,\ r^3,\ r^4,\ r^5\}.$$

The subgroup $\langle r \rangle$ is referred to as the *subgroup generated by r*. In general, the subgroup of a group G generated by a single element x of G is written

$$\langle x \rangle = \{x^n : n \in \mathbb{Z}\}.$$

Exercise 2.4

Determine the elements of the following subgroups of D_6:

(a) $\langle s \rangle$;

(b) $\langle r^i \rangle$, $\quad i = 2, \ldots, 5$;

(c) $\langle r^n s \rangle$, $\quad n = 1, \ldots, 5$.

The subgroups found above, together with $\langle e \rangle$, constitute all the *cyclic* subgroups of D_6: i.e. the complete set of cyclic subgroups of D_6 is

$$\langle e \rangle,\quad \langle r \rangle = \langle r^5 \rangle,\quad \langle r^2 \rangle = \langle r^4 \rangle,\quad \langle r^3 \rangle,\quad \langle s \rangle,\quad \langle rs \rangle,\quad \langle r^2 s \rangle,\quad \langle r^3 s \rangle,\quad \langle r^4 s \rangle,\quad \langle r^5 s \rangle.$$

They do not, however, constitute all the *subgroups* of D_6; they do not, for example, include D_6 itself (because D_6 is not cyclic). We shall return to finding the remaining subgroups of D_6 in the next section of this unit.

Before we move on to our next example, we pause briefly for a definition.

Definition 2.2 *Order of a group element*

If the cyclic subgroup generated by an element x of a group G is finite, then the number of elements in that finite cyclic subgroup is the **order** of the element x. If the cyclic subgroup generated by x is infinite, then the element x is said to have **infinite order**.

We do *not* usually say that x has 'order infinity'.

Thus the orders of the elements s, r, r^2, r^3, r^4, r^5 and $r^n s$ (for $n = 1, \ldots, 5$) of D_6 are 2, 6, 3, 2, 3, 6 and 2 respectively.

Our next example illustrates a general point about groups of symmetries. Consider the frieze in Figure 2.1, which is a modified version of the plain rectangular frieze discussed in Section 1.

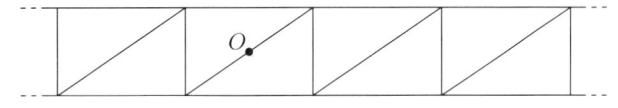

Figure 2.1

Some of the symmetries of the plain rectangular frieze (i.e. some of the elements of group E_1) also preserve the diagonal line that has been added to each rectangle. Let H be the subset of E_1 which consists of symmetries of the plain rectangular frieze that also preserve the new frieze.

There are no symmetries of the new frieze that are not also symmetries of the plain rectangular one.

Exercise 2.5

Describe, geometrically, the elements of H.

You may suspect, quite correctly, that we have introduced H because it is not only a subset but also a subgroup of E_1.

One way of proving that this is so is to turn the description from the last solution into an algebraic one. Since there can be no reflection in H, we are left with symmetries generated by r from $\Gamma(\square)$ and the translation $t[\mathbf{a}]$. Thus an algebraic description of H is

$$H = \left\{ (t[\mathbf{a}])^m r^n : m \in \mathbb{Z},\ n = 0, 1;\ r\, t[\mathbf{a}] = (t[\mathbf{a}])^{-1} r,\ r^2 = e \right\}.$$

With this description, we could check the subgroup axioms, one by one.

However, it is worth considering a more geometric approach, because many of the friezes and tilings considered in the Geometry stream of this course can be obtained by 'decorating' a more basic version, in the same way that the frieze in Figure 2.1 can be obtained by 'decorating' (i.e. adding diagonal lines to) a plain rectangular frieze.

Firstly, closure. Suppose that x and y are two elements of H. Then both are isometries preserving the new frieze. The composite xy is also an isometry. Further, xy means y followed by x; applying y maps the frieze to itself, so applying x to the result does likewise. Thus xy also preserves the frieze.

The argument here is actually a very general one, which shows that if two functions separately preserve some property, then so does the composite function (provided that the composite can be defined).

Next, identity. The identity, e, of the group E_1 clearly preserves the new frieze and so belongs to H.

Lastly, inverses. This is, possibly, the trickiest axiom to check. Suppose that x is an element of H. We know that x has an inverse, x^{-1}, in E_1. What we need to show is that x^{-1} is also in H. We know that

$$x^{-1} x = e$$

and that both x and e preserve the new frieze. For convenience, let us label the new frieze F. Stated algebraically, we know that

$$x(F) = F \quad \text{and} \quad e(F) = F.$$

So

$$\begin{aligned}
& (x^{-1} x)(F) = e(F) = F \\
\Rightarrow\ & x^{-1}(x(F)) = F \\
\Rightarrow\ & x^{-1}(F) = F.
\end{aligned}$$

That is, x^{-1} preserves the frieze and so belongs to H.

That completes the proof that H is a subgroup of E_1.

It is tempting to believe that adding features to a geometric object in $\mathbb{R}^2$ will always lead to a subgroup of the group of symmetries of the original object. This is not true, however, as the following example shows.

The object in Figure 2.2(a) has only the identity element in its symmetry group. If we add a 'half-diagonal', as shown in Figure 2.2(b), we obtain an object with a *larger* symmetry group, namely $\Gamma(\square)$.

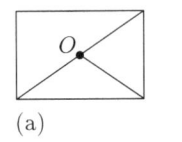

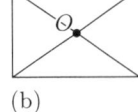

Figure 2.2

Nevertheless, in a number of important cases, adding features does produce a subgroup.

Exercise 2.6

Find the subgroup of $\Gamma(\square)$ which is the symmetry group of the rectangle with a single diagonal in Figure 2.3.

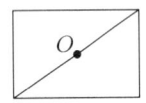

Figure 2.3

Exercise 2.7

Find the subgroup of E_1 which is the symmetry group of the frieze in Figure 2.4.

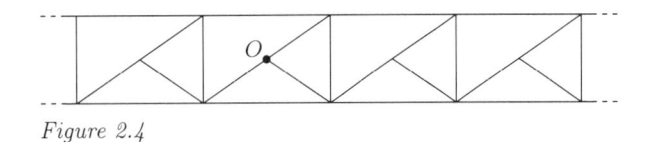

You may find it easier to describe the elements of the subgroup geometrically rather than algebraically.

Figure 2.4

Exercise 2.8

Find the subgroup of D_6 which is the symmetry group of the hexagon with inscribed triangle in Figure 2.5.

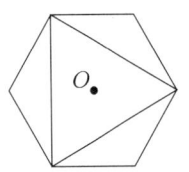

Figure 2.5

You may well have met the subgroup found in the solution to Exercise 2.8 before, possibly labelled as S_3, the group of all permutations of 3 symbols. Since the group is also, effectively, the group of symmetries of an equilateral triangle, it is also perfectly reasonable to call the group D_3. The connection can be seen if you label the corners of the inscribed triangle in Figure 2.5 with 1, 2 and 3. Both S_3 and D_3 are in general use for this group, with S_3 being rather more common.

3 GENERATING SUBGROUPS

In this section we review the examples of subgroups that were discussed in the last section, this time concentrating on the idea of generating subgroups by some subset of the group. The underlying idea is a generalization of the concept of a cyclic subgroup, i.e. a subgroup generated by a single element. We shall also complete the task of finding all the subgroups of D_6.

In Section 2, we found all the cyclic subgroups of D_6 by considering all combinations of copies of each element of D_6 in turn. For example, we found that the rotation r generates the cyclic subgroup

$$\langle r \rangle = \{e, r, r^2, r^3, r^4, r^5\}.$$

In the next exercise, we ask you to see what happens if we extend this process of generation by finding all combinations of copies of *two* distinct elements of D_6.

Note that here we use the phrase 'all combinations of copies of' rather than the phrase 'all integer powers of', which we used in Section 2. The reason is that the new phrase is more generally applicable.

Exercise 3.1

Find the set consisting of all the elements of D_6 produced by combining copies of r^2 and s. Is this set a subgroup of D_6?

In an extension of the notation for cyclic subgroups, we denote the subgroup found in Exercise 3.1 by $\langle r^2, s \rangle$ and refer to it as the *subgroup generated by* r^2 *and* s.

Before we go on to see how we can use the idea of generating subgroups in order to find all the subgroups of D_6, we shall look at another example in order to clarify what is meant by generating a subgroup. The example concerns the group E_1 of symmetries of a plain rectangular frieze.

Exercise 3.2

Describe, geometrically or algebraically, the subset of E_1 obtained by taking all combinations of the translation $t[\mathbf{a}]$ with itself. Is this set a subgroup of E_1?

The example in Exercise 3.2 shows that a little care is needed even for an informal description of what is meant by the following phrase:

'the subgroup generated by...'

Informally, we define the subgroup generated by $x, y, \ldots$ to be the set obtained by forming all possible combinations of copies of $x, y, \ldots$ *and their inverses*. The inclusion of the inverses is quite crucial in ensuring that we actually obtain a subgroup, as it is this that ensures that the identity and inverses axioms for subgroups are satisfied.

A formal definition will be given in *Unit IB4* but is inappropriate here.

Let us now return to the problem of finding all the subgroups of D_6. In the last section, we found all the *cyclic* subgroups of D_6:

$$\langle e \rangle = \{e\};$$
$$\langle r \rangle = \langle r^5 \rangle$$
$$= \{e, r, r^2, r^3, r^4, r^5\};$$
$$\langle r^2 \rangle = \langle r^4 \rangle$$
$$= \{e, r^2, r^4\};$$
$$\langle r^3 \rangle = \{e, r^3\};$$
$$\langle r^n s \rangle = \{e, r^n s\}, \quad n = 0, \ldots, 5.$$

The next set of exercises asks you to find all the remaining subgroups of D_6.

Exercise 3.3

Find all the subgroups $\langle r^m, s \rangle$ for $m = 1, \ldots, 5$.

Exercise 3.4

Explain why $\langle r, r^n s \rangle$ is the whole of D_6, for $n = 0, \ldots, 5$.

Exercise 3.5

(a) Find all the subgroups $\langle r^2, r^n s \rangle$ for $n = 0, 1, \ldots, 5$.
(b) Find all the subgroups $\langle r^3, r^n s \rangle$ for $n = 0, 1, \ldots, 5$.

Exercise 3.6

Explain why each subgroup $\langle r^m, r^n s \rangle$, for $m, n = 0, \ldots, 5$, is one of the subgroups already found.

Exercise 3.7

Explain why each subgroup $\langle r^m s, r^n s \rangle$, for $m, n = 0, \ldots, 5$, is one of the subgroups already found.

As a result of the exercises above, we can now list all the subgroups of D_6 generated by one or two of its elements:

$$\langle e \rangle = \{e\};$$
$$\langle r \rangle = \langle r^5 \rangle$$
$$= \{e, r, r^2, r^3, r^4, r^5\};$$
$$\langle r^2 \rangle = \langle r^4 \rangle$$
$$= \{e, r^2, r^4\};$$
$$\langle r^3 \rangle = \{e, r^3\};$$
$$\langle r^n s \rangle = \{e, r^n s\}, \quad n = 0, \ldots, 5;$$
$$\langle r, r^n s \rangle = D_6, \quad n = 0, \ldots, 5,$$
$$= \{e, r, r^2, r^3, r^4, r^5, s, rs, r^2 s, r^3 s, r^4 s, r^5 s\};$$
$$\langle r^2, s \rangle = \langle r^2, r^2 s \rangle = \langle r^2, r^4 s \rangle$$
$$= \{e, r^2, r^4, s, r^2 s, r^4 s\};$$
$$\langle r^2, rs \rangle = \langle r^2, r^3 s \rangle = \langle r^2, r^5 s \rangle$$
$$= \{e, r^2, r^4, rs, r^3 s, r^5 s\};$$
$$\langle r^3, s \rangle = \langle r^3, r^3 s \rangle$$
$$= \{e, r^3, s, r^3 s\};$$
$$\langle r^3, rs \rangle = \langle r^3, r^4 s \rangle$$
$$= \{e, r^3, rs, r^4 s\};$$
$$\langle r^3, r^2 s \rangle = \langle r^3, r^5 s \rangle$$
$$= \{e, r^3, r^2 s, r^5 s\}.$$

Because this list is so long, and because of the feel we have for the way elements of D_6 combine, it is reasonable to consider that there may be no other subgroups of D_6. We could check this by adding a third generator to each of the two-generator subgroups and proceeding, as in Exercises 3.3–3.7, to see whether this extra generator gives us any new subgroups. However, a better approach is to try to show that *any* subgroup H of D_6 must be one of those in the list.

The easiest argument is basically geometric. Either H contains a reflection or it does not. Suppose first that it does not. Then H contains only rotations and, being non-empty (by the identity axiom), either must be just $\{e\}$ or must contain some r^m and all its powers. Thus, in this case, we have

$$H = \langle r^m \rangle, \quad \text{for some } m = 0, \ldots, 5.$$

Now suppose that H contains a reflection. There are now two sub-cases to consider. The first is that H contains no non-trivial rotation. In this case, H cannot contain a second reflection, because the combination of two different reflections is a non-trivial rotation. Thus, in this sub-case, we have

H must contain the trivial rotation e, by the identity axiom.

$$H = \langle r^n s \rangle = \{e, r^n s\}, \quad \text{for some } n = 0, \ldots, 5.$$

The second sub-case is where H does contain a non-trivial rotation r^m. It follows that H contains the subgroup generated by the rotation and the reflection, that is we have

$$\langle r^m, r^n s \rangle \subseteq H \subseteq D_6.$$

Inspecting the list of subgroups of the form

$$\langle r^m, r^n s \rangle,$$

we see that they have order 4, 6 or 12.

The order of a group is the number of elements it contains.

If the order of $\langle r^m, r^n s \rangle$ is 12, then

$$H = D_6.$$

If the order of $\langle r^m, r^n s \rangle$ is 4, then by Lagrange's Theorem the order of H is divisible by 4 and divides 12. Thus the only possibilities for the order of H in this case are 4 and 12. If the order of H is 4, then, because

$$\langle r^m, r^n s \rangle \subseteq H,$$

we have

$$H = \langle r^m, r^n s \rangle.$$

Lagrange's Theorem says that the order of a subgroup of a finite group divides the order of the group. We assume that you have met this theorem in your previous studies. It is revised in Unit IB4.

If the order of H is 12 then

$$H = D_6.$$

A similar argument applies to the case where the order of $\langle r^m, r^n s \rangle$ is 6, giving that either $H = \langle r^m, r^n s \rangle$ or $H = D_6$.

Now that we know that every subgroup of D_6 has been found, we know that there is no need to consider more than two generators.

The work above shows that, even in the case of a relatively small finite group such as D_6, the task of finding all subgroups can be lengthy and quite laborious, even with the help of Lagrange's Theorem. Some of the work can be eased, but only at the expense of proving some general theorems about how many subgroups of particular orders a finite group can have.

Fortunately, in the case of the groups that arise in the Geometry stream of this course, we shall be more concerned with the *existence* of specific types of subgroups of groups of symmetries of friezes and tilings, rather than with finding all subgroups. With this in mind, we shall, later in the Groups stream of the course, prove some theorems about the existence of subgroups.

4 CYCLIC GROUPS

This is a fairly short section which looks specifically at cyclic groups, that is groups which can be generated by a single element. We shall also consider subgroups of cyclic groups.

4.1 Definition and examples

Definition 4.1 Cyclic group

A group G is a **cyclic group** if there is an element g in G such that
$$G = \langle g \rangle.$$

The definition means that every element of a cyclic group G can be expressed as an integer power of some element g in G.

One example of a cyclic group is the set of non-zero integers
$$\{1, \ldots, 4\}$$
under the operation of multiplication modulo 5, which we denote by $\mathbb{Z}_5^*$. This is cyclic and generated by 2 because, working modulo 5,
$$2^0 = 1, \quad 2^1 = 2, \quad 2^2 = 4, \quad 2^3 = 3, \quad 2^4 = 1.$$

In fact, when p is any *prime* number, the set
$$\mathbb{Z}_p^* = \{1, \ldots, p-1\},$$
under multiplication modulo p, is a cyclic group, though we shall not prove this here.

Exercise 4.1

Show that 3 also generates $\mathbb{Z}_5^*$.

Exercise 4.2

Show that the group of matrices (under matrix multiplication) defined by
$$\left\{ \begin{bmatrix} 1 & a \\ 0 & 1 \end{bmatrix} : a \in \mathbb{Z} \right\}$$
is cyclic by finding a generator.

We showed that this set is a group in Exercise 1.3.

The solution to Exercise 4.1 shows that a cyclic group may have more than one possible choice of generator.

Cyclic groups have a number of properties not shared by groups in general. An important one is that a cyclic group must be Abelian.

A group G is Abelian if $xy = yx$ for all pairs x, y of elements in G.

This is so because, if G is cyclic and generated by g, then any two elements x, y in G will be of the form
$$x = g^m, \quad y = g^n,$$
for some integers m and n. Then
$$\begin{aligned} xy &= g^m g^n \\ &= g^{m+n} \\ &= g^{n+m} \\ &= g^n g^m \\ &= yx. \end{aligned}$$

We shall see later that cyclic groups also have subgroup properties that groups in general do not have.

A number of *additive* groups (i.e. those whose binary operation is addition) are cyclic. The most obvious example of these is the group of integers, $\mathbb{Z}$, under addition. Before we consider such groups, a few words are in order about notation for groups written additively.

When n is a positive integer, the general symbol x^n means n copies of x combined using the group operation. In additive notation this becomes

$$\overbrace{x + \cdots + x}^{n \text{ copies}}$$

which we write as nx rather than x^n. Similarly, when n is a negative integer, x^n means $|n|$ copies of x^{-1} (the inverse of x) combined using the group operation. In additive notation this becomes

$$\overbrace{(-x) + \cdots + (-x)}^{|n| \text{ copies}} = |n|(-x) = nx.$$

Remember that $|n| = -n$ when n is a negative integer.

Finally, x^0 is the identity of the group; in additive notation the identity is usually written as 0, so we write

$$0x = 0.$$

Note here that the two appearances of 0 have different meanings: the first is the integer 0, the second is the identity of the group being considered.

If we want to emphasize that we are calculating a 'power' in an additive group, we shall write $n \cdot x$ instead of just nx.

We can now see why $\mathbb{Z}$, under addition, is cyclic, generated by 1, because if n is any integer, we have

$$n = n \cdot 1.$$

Exercise 4.3

(a) Show that the set of integers

$$\{0, \ldots, 4\}$$

under addition modulo 5, denoted by $\mathbb{Z}_5$, is a cyclic group.

(b) How many different generators has $\mathbb{Z}_5$?

Although only $\mathbb{Z}_5$ is considered in the above exercise, we hope that you can see that, in general, the set of integers

$$\{0, \ldots, n-1\}$$

under addition modulo n, denoted by $\mathbb{Z}_n$, is a cyclic group generated by 1. However, it is not true in general that any non-zero element of $\mathbb{Z}_n$ will generate $\mathbb{Z}_n$, as we shall see in the next subsection.

Among the examples that you have met so far, there are some non-cyclic groups. For example, D_6 is not cyclic.

Exercise 4.4

Why is D_6 not cyclic?

In the examples of cyclic groups discussed above, two are of particular importance: $\mathbb{Z}$ and $\mathbb{Z}_n$, both under addition. Their importance stems from the fact that they are prototypes for *all* cyclic groups in a sense that we shall outline here.

First, we make the 'obvious' remark that a cyclic group is either finite or infinite and deal with the two cases separately.

This is true of any group.

Suppose that G is an *infinite* cyclic group (written multiplicatively) generated by the element g. That means that every element is g^n for some integer n. We can set up a mapping ϕ from $\mathbb{Z}$ to G as follows:

$$\phi : n \mapsto g^n.$$

It turns out that ϕ is an isomorphism from $\mathbb{Z}$ to G, which means that $\mathbb{Z}$ and G are essentially the same.

Isomorphisms are discussed more formally and in more detail in Unit IB4.

The two properties that make ϕ an *isomorphism* from $\mathbb{Z}$ to G are, first, that the operations in the two groups should correspond properly under ϕ and, second, that ϕ is one–one and onto (i.e. is a *bijection*).

It is quite straightforward to show that the first property holds, since using the rule for indices

$$g^{m+n} = g^m g^n$$

we have

$$\phi(m+n) = g^{m+n} = g^m g^n = \phi(m) + \phi(n).$$

Thus the operations in the two groups correspond under ϕ.

We ask you to prove that the second property holds in the following two exercises.

Exercise 4.5

Show that ϕ is onto.

Hint Use the fact that g generates G.

Exercise 4.6

Show that ϕ is one–one.

Hint Use the fact that G is infinite.

Thus we have proved that every infinite cyclic group is isomorphic to $\mathbb{Z}$.

In a similar way, it can be proved that any *finite* cyclic group with n elements is isomorphic to $\mathbb{Z}_n$. The proof requires some of the properties of the integers that will be discussed in *Unit GR1*, and so is deferred to *Unit GR2*.

These results mean that, to prove a statement for all cyclic groups, it is sufficient to prove it for $\mathbb{Z}$ and $\mathbb{Z}_n$.

4.2 Subgroups of cyclic groups

When we were looking for subgroups of groups earlier in this unit, we started by finding cyclic subgroups. We took a particular element and found the cyclic subgroup that it generated. Thus we saw that non-cyclic groups have cyclic subgroups.

The converse is false, however: cyclic groups can have *only* cyclic subgroups. We cannot give a complete proof here, since it requires some of the properties of the integers that will be discussed in *Unit GR1*. For now, we shall just look at a simple case to show that the result is believable.

Consider $\mathbb{Z}_8$, the set of integers modulo 8 under addition. This is a cyclic group generated by, for example, 1. The following exercises ask you to investigate $\mathbb{Z}_8$.

Exercise 4.7

For each element g in $\mathbb{Z}_8$, write down the cyclic subgroup generated by g.

Exercise 4.8

Show that any two elements of $\mathbb{Z}_8$ generate a cyclic subgroup.

Hint Most of the choices of pairs can be disposed of very quickly.

A case-by-case analysis would show that all subgroups of $\mathbb{Z}_8$ are cyclic. This case-by-case style of argument, however, does not generalize. A proof which does generalize is based on properties of the integers and will be discussed in *Unit GR2*.

Before we leave cyclic groups for the time being, it is worth noting that there are two 'standard' notations for a cyclic group with n elements, depending on the context: algebraic or geometric.

When we are discussing algebraic properties of groups, we will usually use

$$\mathbb{Z}_n = \{0, \ldots, n-1\}$$

under addition modulo n as our model.

In a geometric context, we are likely to take as a model the group

$$C_n = \langle r[2\pi/n] \rangle,$$

that is the group generated by the rotation $r[2\pi/n]$, under composition of functions.

5 GROUP ACTIONS

This section introduces group actions, and provides two examples of group actions that will be particularly relevant to the course.

The treatment here is slightly more formal than the one you may have met in a previous course.

The idea of a group 'acting on' a set of objects has been a feature of many of our examples of groups. We have groups of symmetries which act on the points making up a geometric figure, groups of permutations acting on sets like $\{1, \ldots, n\}$, and so on. In each case the elements of the group are (or behave as if they are) *functions* from a set to itself. In the cases that you have met, the function associated with each group element has special properties: it is both one–one and onto (i.e. the function is a bijection).

The formal definition of a *group action* generalizes these ideas. There are three parts to a group action: a group G, a set X on which the group *acts* and an association between the group and the set of bijections from the set X to itself.

There are therefore two operations involved: the group operation and composition of the bijections. The central feature of a group action is that combination of elements of the group corresponds exactly to composition of the corresponding bijections.

It is convenient to introduce a notation for the set of bijections from X to itself. By analogy with symmetry groups, we denote this set of bijections by $\Gamma(X)$. It is not difficult to prove that $\Gamma(X)$ is a group under the operation of composition, although we do not do so here.

The proof is an application of the fact that the set of functions preserving something is always a group.

Associating a bijection in $\Gamma(X)$ with each element of G means that we must have a function from G to $\Gamma(X)$. We call this function ϕ, and we denote the image of g under ϕ by ϕ_g. For each $g \in G$, we emphasize that

$$\phi_g : X \to X$$

is a bijection.

We are now in a position to define a group action.

> **Definition 5.1 Group action**
>
> A **group action** consists of a group G, a set X and a mapping
>
> $$\phi : G \to \Gamma(X)$$
> $$g \mapsto \phi_g$$
>
> from G to the group $\Gamma(X)$ of bijections from X to X, such that
>
> $$\phi_{gh} = \phi_g \circ \phi_h, \quad \text{for all } g, h \in G.$$

You may well recognize that this definition can be rephrased as saying that ϕ is a homomorphism from the group G to the group $\Gamma(X)$. As a consequence, because homomorphisms map identities to identities, the bijection ϕ_e, corresponding to the identity of G, is the identity mapping from X to X.

We are assuming some familiarity with the definition of a homomorphism. The idea will be developed in Unit IB4.

In practice we do not usually use the formal terminology and notation of the above definition. If we want to say what happens to an element x of the set X under the action of an element g of the group G, we ought to refer to

'the image of x under ϕ_g',

that is to

$$\phi_g(x).$$

Or, more fully, to 'the image of x under the bijection corresponding to g'.

It is much more in keeping with the examples of group actions to modify the terminology and refer to

'the image of x under g'

and to modify the notation and write

$$g \wedge x \quad \text{instead of} \quad \phi_g(x).$$

We shall in general prefer to use the less formal $\wedge$ (wedge) notation and corresponding terminology. However, the formal terminology and notation can prove useful in drawing attention to the similarities between group actions and other groups of functions.

We can use our formal definition of a group action to deduce the following properties of a group action in terms of the $\wedge$ notation.

> **Lemma 5.1**
>
> For an action of the group G on the set X, we have:
> (a) $g \wedge x \in X$, for all $g \in G$, $x \in X$;
> (b) $e \wedge x = x$, for all $x \in X$, where e is the identity element of G;
> (c) $(gh) \wedge x = g \wedge (h \wedge x)$, for all $g, h \in G$ and $x \in X$.

Proof

The proof follows from the comments above.

(a) This just states that each ϕ_g is a function from X to X.

(b) This follows from our remark that ϕ_e is the identity bijection on X.

(c) This is simply the translation of

$$\phi_{gh} = \phi_g \circ \phi_h$$

into $\wedge$ notation. ∎

The conditions in Lemma 5.1 can be considered to provide an alternative definition of a group action. To see this, suppose we are given $\wedge$, X and G with $\wedge$ satisfying the three conditions in Lemma 5.1. Then we can define a group action of G on X by taking

$$\phi_g(x) = g \wedge x$$

for all $x \in X$ and $g \in G$. It is fairly easy to check that the conditions on $\wedge$ ensure that this *does* define a group action. Condition (a) says that ϕ_g is a function from X to X and Condition (c) gives the homomorphism property. What is missing is the bijection property. However, this follows from the other conditions as follows.

Firstly, ϕ_g is one–one because, for $x, y \in X$,

$$\begin{aligned}
& \phi_g(x) = \phi_g(y) \\
\Rightarrow\quad & g \wedge x = g \wedge y \\
\Rightarrow\quad & g^{-1} \wedge (g \wedge x) = g^{-1} \wedge (g \wedge y) \quad \text{(by Condition (a))} \\
\Rightarrow\quad & (g^{-1}g) \wedge x = (g^{-1}g) \wedge y \quad \text{(by Condition (c))} \\
\Rightarrow\quad & e \wedge x = e \wedge y \\
\Rightarrow\quad & x = y \quad \text{(by Condition (b))}.
\end{aligned}$$

Secondly, ϕ_g is onto since, if $x \in X$,

$$\begin{aligned}
x &= e \wedge x \quad \text{(by Condition (b))} \\
&= (gg^{-1}) \wedge x \\
&= g \wedge (g^{-1} \wedge x) \quad \text{(by Condition (c))} \\
&= \phi_g(y), \quad \text{where } y = g^{-1} \wedge x \in X \text{ (by Condition (a))}.
\end{aligned}$$

It is high time for some examples of group actions.

The group E_2 of symmetries of a plain rectangular tiling certainly acts on the set of all points of the tiling. However, there are other sets on which it acts as well.

For example, we can take X to be the set consisting of all tiles, edges and vertices of the tiling — that is, the set of all *parts* of the tiling. The group action requirements then follow from the definition of E_2 as the set of symmetries of the tiling, as the following exercise illustrates.

Exercise 5.1

Describe the edge $t[\mathbf{a}] \wedge E$, where E is the edge marked in Figure 5.1.

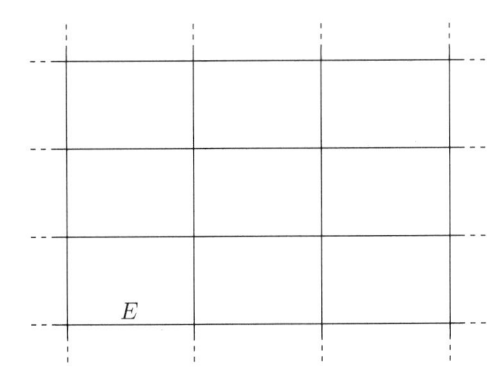

Figure 5.1

A second example that we want to explore is rather more complicated, not because the idea is harder but because the set X being acted upon is closely involved with the group doing the acting.

The group in this example is D_6. The set X is the set of elements of D_6. An element of D_6 is to act on an element of X by the following operation. If g is in D_6 and x is in X then

$$g \wedge x = gxg^{-1}.$$

You may need to refer back to our description of D_6 in terms of r and s.

The element gxg^{-1} is called the *conjugate* of x by g.

For example, r is an element of D_6 and rs is an element of X, so

$$\begin{aligned}
r \wedge (rs) &= r(rs)r^{-1} \\
&= rrsr^5 \quad \left(\text{using } r^{-1} = r^5\right) \\
&= r^2 srrrrr \\
&= r^2 r^5 r^5 r^5 r^5 r^5 s \quad \left(\text{using } sr = r^5 s \text{ five times}\right) \\
&= r^2 r^{-5} s \quad \left(\text{using } r^5 = r^{-1}\right) \\
&= r^{-3} s \\
&= r^3 s \quad \left(\text{using } r^6 = e\right).
\end{aligned}$$

The following exercises concern the action of D_6 on itself described above.

Exercise 5.2

Calculate:

(a) $s \wedge r$;

(b) $r^n \wedge r, \quad n = 0, \ldots, 5$;

(c) $r^n s \wedge r, \quad n = 0, \ldots, 5$.

Exercise 5.3

Calculate:

(a) $s \wedge s$;

(b) $r^n \wedge s, \quad n = 0, \ldots, 5$;

(c) $r^n s \wedge s, \quad n = 0, \ldots, 5$.

Exercise 5.4

Calculate:

(a) $s \wedge rs$;

(b) $r^n \wedge rs, \quad n = 0, \ldots, 5$;

(c) $r^n s \wedge rs, \quad n = 0, \ldots, 5$.

This last set of exercises illustrates one of the two concepts associated with group actions that we want to discuss here, the concept of an *orbit*.

> **Definition 5.2 Orbit**
>
> If a group G acts on a set X and x is an element of X, then the **orbit** of x under G is the set of elements of X obtained by acting on x with the elements of G. It is denoted by
>
> $$\operatorname{Orb}(x) = \{g \wedge x : g \in G\}.$$

In our first example of a group action, using E_2, the orbit of the edge E in Figure 5.1 is the set of all long edges of the tiles. This is because the image of a long edge can only be a long edge and because any such edge can be mapped to any other.

Exercise 5.5

Using the results of Exercises 5.2–5.4, determine the orbits of all the elements of D_6 under the action of D_6 defined by $g \wedge x = gxg^{-1}$.

Exercise 5.6

Determine the orbits of all the elements of the set X of all parts (i.e. tiles, edges and vertices) of a plain rectangular tiling under the action of the group E_2 of symmetries of the tiling.

In the solutions to Exercises 5.5 and 5.6, you may have noticed that the orbits *partition* the set being acted upon, that is the orbits fill out the whole of the set being acted upon and do not overlap (unless they coincide). For example, in the action of D_6, the six sets we obtained as orbits were

This suggests that 'is in the same orbit as' defines an equivalence relation on the set X.

$$\{e\},$$
$$\{r, r^5\},$$
$$\{r^2, r^4\},$$
$$\{r^3\},$$
$$\{s, r^2s, r^4s\},$$
$$\{rs, r^3s, r^5s\},$$

and these six sets partition D_6.

We shall generalize this property to all group actions later in the Groups stream. The property is useful for proofs about finite groups, because it tells us that the sum of the number of elements in each orbit must be the total number of elements in the set.

Now, the orbit of an element of X is a subset of the set X. The final idea that we want to look at in this unit leads to a subset of the group G, rather than of the set X.

Definition 5.3 Stabilizer

If a group G acts on a set X and x is an element of X, then the **stabilizer** of x under G is the set of elements of G which fix x. It is denoted by
$$\text{Stab}(x) = \{g : g \in G, \ g \wedge x = x\}.$$

For example, in the E_2 example of a group action, the elements of E_2 which fix the edge labelled E in Figure 5.1 are:

reflection in the line through E;

rotation through π about the centre of E;

reflection in a line through the centre of E, perpendicular to E;

the identity.

These four isometries form not only a sub*set* of E_2 but also a sub*group* of E_2. This is no coincidence, because, in general, $\text{Stab}(x)$ is a subgroup of G for any x in a set X acted on by a group G.

This subgroup of E_2 is isomorphic to $\Gamma(\square)$.

Exercise 5.7

Show that, if G acts on X and $x \in X$, then $\text{Stab}(x)$ is a subgroup of G.

We repeat: *orbits* are *subsets* of X; *stabilizers* are *subgroups* of G.

Exercise 5.8

Determine the stabilizers of the following elements of D_6 under the action of D_6 defined by $g \wedge x = gxg^{-1}$:

(a) e;

(b) r;

(c) s.

Hint Use your solutions to Exercises 5.2–5.5.

To conclude this discussion, we note that the action of D_6 on itself that we defined above is not the only one possible. We shall consider some others later in the course. Also, we note that the definition

$$g \wedge x = gxg^{-1}$$

can be generalized to define an action of *any* group on itself.

SOLUTIONS TO THE EXERCISES

Solution 1.1

We check the axioms in turn.

Closure The sum of two integers, positive or negative, is another integer.

Associativity From what we know about addition, we can assume that $(x+y)+z = x+(y+z)$ for all choices of integers x, y and z.

Identity The integer 0 is the identity element for addition: for any integer x
$$x + 0 = x = 0 + x.$$

Inverses The inverse of an integer x is the integer $-x$ because
$$x + (-x) = 0 = (-x) + x$$
and 0 is the identity element of $\mathbb{Z}$.

Solution 1.2

Given associativity, we need only check the other three axioms.

Closure If $\mathbf{A}$ and $\mathbf{B}$ are two 2×2 matrices with real entries, the process of matrix multiplication guarantees that the product $\mathbf{AB}$ is 2×2 with real entries. This is not quite enough, since we must also show that the determinant of $\mathbf{AB}$ is non-zero. It is probably easiest to tackle this by first showing that, in general,
$$\det(\mathbf{AB}) = \det(\mathbf{A})\det(\mathbf{B})$$
for any 2×2 matrices $\mathbf{A}$ and $\mathbf{B}$. So, suppose that
$$\mathbf{A} = \begin{bmatrix} a & c \\ b & d \end{bmatrix} \quad \text{and} \quad \mathbf{B} = \begin{bmatrix} e & g \\ f & h \end{bmatrix}.$$
Then
$$\mathbf{AB} = \begin{bmatrix} ae+cf & ag+ch \\ be+df & bg+dh \end{bmatrix}$$
and so
$$\begin{aligned}\det(\mathbf{AB}) &= (ae+cf)(bg+dh) - (ag+ch)(be+df) \\ &= aebg + aedh + cfbg + cfdh - agbe - agdf - chbe - chdf \\ &= aedh + cfbg - agdf - chbe \\ &= ad(eh - gf) - bc(eh - gf) \\ &= (ad - bc)(eh - gf) \\ &= \det(\mathbf{A})\det(\mathbf{B}).\end{aligned}$$
So, if $\mathbf{A}$ and $\mathbf{B}$ both have non-zero determinants, so has $\mathbf{AB}$. This completes the closure argument.

Identity The matrix
$$\mathbf{I} = \begin{bmatrix} 1 & 0 \\ 0 & 1 \end{bmatrix}$$
acts as an identity for matrix multiplication *and* has determinant 1, which is non-zero.

Inverses The standard procedure for finding an inverse for $\mathbf{A}$ as
$$\mathbf{A}^{-1} = \frac{1}{ad - bc}\begin{bmatrix} d & -c \\ -b & a \end{bmatrix}$$
can be used whenever $\det(\mathbf{A}) = ad - bc \neq 0$, and produces a 2×2 matrix with real entries. The inverse $\mathbf{A}^{-1}$ also has non-zero determinant since
$$\det(\mathbf{A})\det(\mathbf{A}^{-1}) = \det(\mathbf{I}) = 1.$$

By the way, this also shows that $\det(\mathbf{A}^{-1}) = 1/\det(\mathbf{A})$.

Solution 1.3

Given associativity, we need only check the other three axioms.

Closure We must take two typical elements and show that the product is of the right form to be in the specified set. So, suppose that a, b are in $\mathbb{Z}$; then

$$\begin{bmatrix} 1 & a \\ 0 & 1 \end{bmatrix} \begin{bmatrix} 1 & b \\ 0 & 1 \end{bmatrix} = \begin{bmatrix} 1 & a+b \\ 0 & 1 \end{bmatrix}.$$

Since $a + b$ is also in $\mathbb{Z}$, the product of the two matrices is also in the set. We have closure.

Identity The matrix

$$\mathbf{I} = \begin{bmatrix} 1 & 0 \\ 0 & 1 \end{bmatrix}$$

acts as an identity for matrix multiplication. However, we need to check that it belongs to the set in question. Since the leading diagonal elements are 1, the bottom left element is 0 and the top right element is an integer (namely 0), $\mathbf{I}$ is in the set.

Inverses Again, the problem is not so much whether

$$\begin{bmatrix} 1 & a \\ 0 & 1 \end{bmatrix}$$

has a multiplicative inverse (its determinant is $1 \neq 0$, so it does), but whether the inverse is present in the specified set. Since the usual way of calculating inverses gives

$$\begin{bmatrix} 1 & a \\ 0 & 1 \end{bmatrix}^{-1} = \begin{bmatrix} 1 & -a \\ 0 & 1 \end{bmatrix},$$

and the inverse has the correct form (since $-a$ is an integer whenever a is), the inverse of each element of the set does indeed belong to the set.

Solution 1.4

The functions compose as follows:

$$1 \mapsto 3 \mapsto 4$$
$$2 \mapsto 5 \mapsto 2$$
$$3 \mapsto 1 \mapsto 3$$
$$4 \mapsto 2 \mapsto 5$$
$$5 \mapsto 4 \mapsto 1$$

Thus

$$\begin{pmatrix} 1 & 2 & 3 & 4 & 5 \\ 3 & 5 & 4 & 1 & 2 \end{pmatrix} \begin{pmatrix} 1 & 2 & 3 & 4 & 5 \\ 3 & 5 & 1 & 2 & 4 \end{pmatrix} = \begin{pmatrix} 1 & 2 & 3 & 4 & 5 \\ 4 & 2 & 3 & 5 & 1 \end{pmatrix}.$$

Because permutations are functions, composition is done from right to left.

Solution 1.5

We start by tracing the image of 1, the image of this image, and so on, until we come back to 1. Then we consider the first number not already used and repeat the process. This gives the following:

$$1 \mapsto 3 \mapsto 4 \mapsto 1$$
$$2 \mapsto 5 \mapsto 2$$

Hence the original permutation can be written in cycle form as

$$(1\,3\,4)(2\,5).$$

Because the cycles in cycle form are disjoint, the order of the cycles does not matter. So, $(1\,3\,4)(2\,5)$ and $(2\,5)(1\,3\,4)$ are equivalent cycle forms for the given permutation.

Solution 1.6

Here the three cycles are not disjoint, so the product is not yet in cycle form. To write it in cycle form, we work 'right to left' and trace what happens to each number and its images in turn. We obtain:

$$1 \mapsto 3 \mapsto 3 \mapsto 1$$
$$2 \mapsto 1 \mapsto 1 \mapsto 6$$
$$6 \mapsto 2 \mapsto 5 \mapsto 5$$
$$5 \mapsto 6 \mapsto 6 \mapsto 3$$
$$3 \mapsto 5 \mapsto 4 \mapsto 4$$
$$4 \mapsto 4 \mapsto 2 \mapsto 2$$

Hence we have

$$(1\,6\,3)(2\,5\,4)(1\,3\,5\,6\,2) = (1)(2\,6\,5\,3\,4) = (2\,6\,5\,3\,4).$$

We usually omit cycles of length one.

Solution 1.7

The image of O must be the centre of one of the rectangles in the frieze. To see why, consider what must happen to the base rectangle under a symmetry of the frieze. Using the 'corners map to corners' argument that we have used before, it must map to a rectangle. Now we can use the distance-preserving property of isometries. The centre O of the base rectangle is equidistant from the four corners of the base rectangle. So, the image of O must be equidistant from the images of the four corners of the base rectangle. This forces the image of O to be the centre of a rectangle.

Therefore the set of possible images of O is the set of centres of rectangles in the frieze.

Solution 1.8

The set of possible images is the set of *all* corners of *all* rectangles in the frieze.

Solution 1.9

The set of possible images is the same as in the previous exercise.

However, both ends of the top side cannot map to the *same* corner because isometries must be one–one (and onto).

Solution 1.10

(a) First $r\,t[\mathbf{a}]$:

Now $(t[\mathbf{a}])^{-1}\,r$:

Thus, $r\,t[\mathbf{a}] = (t[\mathbf{a}])^{-1}\,r$.

(b) First $h\,t[\mathbf{a}]$:

Now $t[\mathbf{a}]\,h$:

Thus, $h\,t[\mathbf{a}] = t[\mathbf{a}]\,h$.

The diagrams show the effect on the base rectangle, first of $r\,t[\mathbf{a}]$, then of $(t[\mathbf{a}])^{-1}\,r$.

(c) First $v\,t[\mathbf{a}]$:

Now $(t[\mathbf{a}])^{-1}\,v$:

Thus $v\,t[\mathbf{a}] = (t[\mathbf{a}])^{-1}\,v$.

Solution 1.11

(a) We have
$$\begin{aligned}(t[\mathbf{a}])^2\,r\,(t[\mathbf{a}])^3\,v &= (t[\mathbf{a}])^2\,r\,t[\mathbf{a}]\,t[\mathbf{a}]\,t[\mathbf{a}]\,v \\ &= (t[\mathbf{a}])^2\,(t[\mathbf{a}])^{-1}\,r\,t[\mathbf{a}]\,t[\mathbf{a}]\,v \\ &= (t[\mathbf{a}])^2\,(t[\mathbf{a}])^{-1}\,(t[\mathbf{a}])^{-1}\,r\,t[\mathbf{a}]\,v \\ &= (t[\mathbf{a}])^2\,(t[\mathbf{a}])^{-1}\,(t[\mathbf{a}])^{-1}\,(t[\mathbf{a}])^{-1}\,(r\,v) \\ &= (t[\mathbf{a}])^{-1}h.\end{aligned}$$

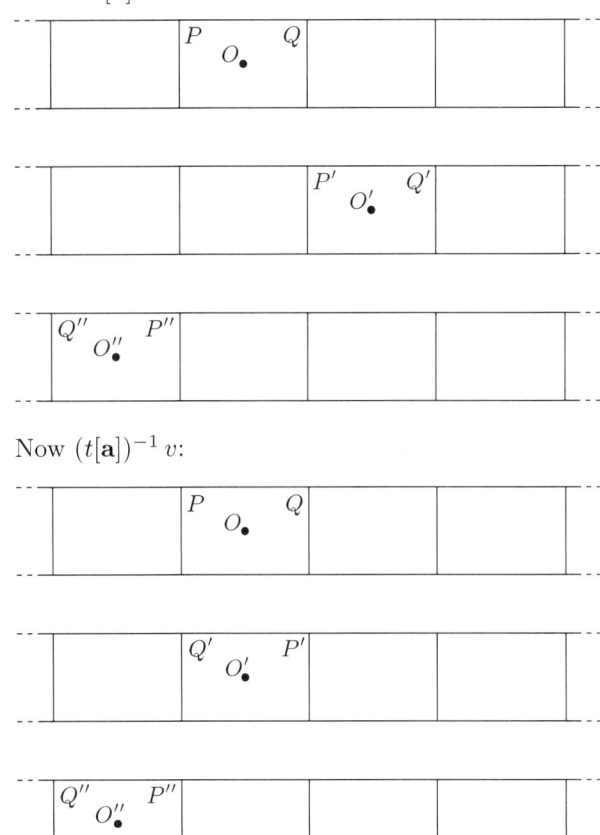

(b) This time
$$\begin{aligned}(t[\mathbf{a}])^3\,v\,(t[\mathbf{a}])^2\,r &= (t[\mathbf{a}])^3\,v\,t[\mathbf{a}]\,t[\mathbf{a}]\,r \\ &= (t[\mathbf{a}])^3\,(t[\mathbf{a}])^{-1}\,v\,t[\mathbf{a}]\,r \\ &= (t[\mathbf{a}])^3\,(t[\mathbf{a}])^{-1}\,(t[\mathbf{a}])^{-1}\,(v\,r) \\ &= t[\mathbf{a}]\,h.\end{aligned}$$

Solution 1.12

By the usual argument, corners must go to corners. Having moved one corner, there are only two places that an adjacent corner can be mapped to: the adjacent corners to the image corner. This gives a maximum of twelve symmetries.

There are six rotations about the centre O: through angles of 0, $\pi/3$, $2\pi/3$, $3\pi/3\,(=\pi)$, $4\pi/3$ and $5\pi/3$ anticlockwise.

We could equally well have included rotation through 2π instead of 0.

There are also six reflections: three in lines joining pairs of opposite corners, three in lines joining midpoints of opposite sides.

This shows that there are exactly twelve symmetries of a regular hexagon.

Solution 1.13

All the rotations are obtained by repeating r a sufficient number of times. Thus the rotations are

$$r, r^2, r^3, r^4, r^5, r^6 = r^0 = e.$$

Solution 1.14

The easiest way to find out which reflection corresponds to each expression of the form $r^n s$ is to keep track of what happens to the corners under each composite transformation. For example, the effect of rs is shown below.

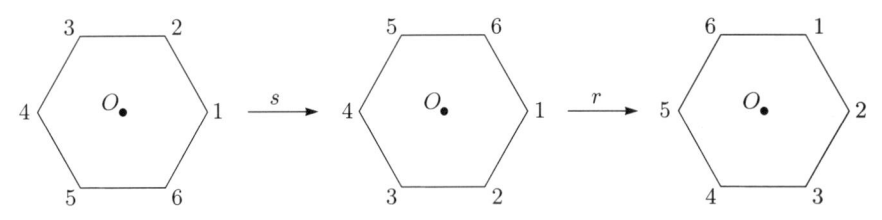

Its effect is therefore reflection in the line joining the midpoints of the opposite sides 12 and 45.

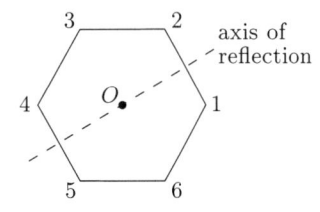

Similar considerations show that

$$rs, r^3 s, r^5 s$$

are the three reflections in the lines joining the midpoints of opposite sides, whereas

$$r^2 s, r^4 s, r^6 s = r^0 s = es = s$$

are the reflections in the lines joining pairs of opposite corners.

Solution 1.15

By experiment, we find that $sr = r^5 s$.

You may have expressed this in the form $r^{-1} s$, which is the same because $r^{-1} = r^5$.

Solution 1.16

The seven results can all be verified by showing what happens to the base rectangle centred at O, in exactly the same way as we did for the plain rectangular frieze.

Solution 2.1

Since, by the identity axiom, each subgroup of $\Gamma(\square)$ must contain the identity element, the only possible candidates for subgroups are:

$$\{e\}, \quad \{e,r\}, \quad \{e,h\}, \quad \{e,v\}, \quad \{e,r,h\}, \quad \{e,r,v\}, \quad \{e,h,v\}, \quad \Gamma(\square).$$

The set $\{e\}$, consisting of just the identity element of G, trivially satisfies the subgroup axioms, as a glance at its Cayley table shows:

$\circ$	e
e	e

For the two-element subsets, the Cayley tables are:

$\circ$	e	r
e	e	r
r	r	e

$\circ$	e	h
e	e	h
h	h	e

$\circ$	e	v
e	e	v
v	v	e

In each case, inspection of these tables shows that the subgroup axioms are satisfied.

None of the three-element subsets satisfies the closure axiom, since

$$rh = hr = v, \quad rv = vr = h, \quad hv = vh = r.$$

The whole group, $\Gamma(\square)$, automatically satisfies the subgroup axioms.

Therefore the five subgroups of $\Gamma(\square)$ are:

$$\{e\}, \quad \{e,r\}, \quad \{e,h\}, \quad \{e,v\}, \quad \Gamma(\square).$$

Solution 2.2

(a) The set R of all rotations in E_1 is not closed. In Section 1 we found that $r\,t[\mathbf{a}]$ was a rotation. So we have

$$r, r\,t[\mathbf{a}] \in R.$$

But their composite is not in R:

$$r\,r\,t[\mathbf{a}] = t[\mathbf{a}] \quad \text{(since } r^2 = e\text{)},$$

and $t[\mathbf{a}]$ is a non-tirvial translation and so is not in R.

(b) We check the axioms in turn.

Closure Typical elements of T are $(t[\mathbf{a}])^m$ and $(t[\mathbf{a}])^n$, for integers m and n. Now

$$(t[\mathbf{a}])^m (t[\mathbf{a}])^n = (t[\mathbf{a}])^{m+n},$$

which is also an integer power of $t[\mathbf{a}]$ and hence is in T.

Identity We defined any element to the power zero to be the identity, so e is indeed an integer power of $t[\mathbf{a}]$ and hence is in T.

Inverses The inverse of a typical element $(t[\mathbf{a}])^n$ of T is $(t[\mathbf{a}])^{-n}$, because

$$(t[\mathbf{a}])^n (t[\mathbf{a}])^{-n} = (t[\mathbf{a}])^0 = e = (t[\mathbf{a}])^{-n} (t[\mathbf{a}])^n.$$

Since the inverse is an integer power of $t[\mathbf{a}]$, it is in T.

Thus T is a subgroup of E_1.

Solution 2.3

For $\{e\}$, the proof given for $\{e\}$ in the case of the group $\Gamma(\square)$ in Solution 2.1 works for any group G.

For G, note that G is a subset of itself and that the group axioms (which G obeys) automatically ensure that G satisfies the subgroup axioms.

Solution 2.4

In this solution, we repeatedly use the relations between the generators of D_6, particularly $r^6 = e$ (which means that all calculations with powers of r are done by reducing the indices modulo 6).

(a) Since $s^2 = e$, the subgroup contains only two elements:
$$\langle s \rangle = \{e, s\}.$$

(b) We deal with the values of i in turn.

For $i = 2$, we have
$$(r^2)^2 = r^4 \quad \text{and} \quad (r^2)^3 = r^6 = e,$$
so
$$\langle r^2 \rangle = \{e, r^2, r^4\}.$$

For $i = 3$, we have
$$(r^3)^2 = r^6 = e,$$
so
$$\langle r^3 \rangle = \{e, r^3\}.$$

For $i = 4$, we have
$$(r^4)^2 = r^8 = r^2 \quad \text{and} \quad (r^4)^3 = r^{12} = r^0 = e,$$
so
$$\langle r^4 \rangle = \{e, r^2, r^4\}. \qquad \text{Note that } \langle r^4 \rangle = \langle r^2 \rangle.$$

Finally, for $i = 5$, we have
$$(r^5)^2 = r^{10} = r^4,$$
$$(r^5)^3 = r^{15} = r^3,$$
$$(r^5)^4 = r^{20} = r^2,$$
$$(r^5)^5 = r^{25} = r^1 = r,$$
$$(r^5)^6 = r^{30} = r^0 = e,$$
so
$$\langle r^5 \rangle = \{e, r, r^2, r^3, r^4, r^5\}. \qquad \text{Note that } \langle r^5 \rangle = \langle r \rangle.$$

(c) Here is a case where remembering the geometric origins of D_6 makes the calculation easier. Each element $r^n s$ is a reflection, so its square must be the identity. Thus
$$\langle r^n s \rangle = \{e, r^n s\}, \quad n = 1, \ldots, 5.$$

The same result can also be obtained algebraically by using
$$sr = r^5 s$$
repeatedly. For example,
$$(r^2 s)^2 = rrsrrs$$
$$= rrr^5 srs$$
$$= rrr^5 r^5 ss$$
$$= r^{12} e$$
$$= ee$$
$$= e.$$

Solution 2.5

First of all, consider the elements of E_1 that fix the base rectangle centred at O, that is the elements of $\Gamma(\square)$. Both of the reflections in $\Gamma(\square)$, h and v, map the marked diagonal to the other diagonal, so cannot be in H. We are left with only the rotation r and the identity e from $\Gamma(\square)$.

The translation $t[\mathbf{a}]$ (and all its integer powers) do preserve the set of diagonals marked.

So the subset H consists of all elements obtainable by combining e and r from $\Gamma(\square)$ with integer powers of $t[\mathbf{a}]$.

Solution 2.6

By an argument similar to the one at the start of Solution 2.5, the subgroup is

$$\{e, r\}.$$

We have seen that this set *is* a subgroup in Exercise 2.1.

Solution 2.7

Arguing about the 'base rectangle' as in Solution 2.5, we now lose r from the subgroup as well. The only surviving member of $\Gamma(\square)$ is e.

However, all translations in E_1 still preserve the new frieze.

The subgroup consists of all the translations of E_1 (including the identity as the trivial translation). Algebraically the subgroup is

$$\{(t[\mathbf{a}])^n : n \in \mathbb{Z}\}.$$

We have see that this set *is* a subgroup in Exercise 2.2.

Solution 2.8

Consider the rotations first. r no longer preserves the figure, but r^2 does. This is because only rotations through multiples of $2\pi/3$ preserve the figure. Thus e, r^2 and r^4 belong to the subgroup.

Now consider the reflections. Only the reflections through corners of the inscribed triangle preserve the figure. These are s, r^2s and r^4s.

Thus the subgroup is

$$\{e, r^2, r^4, s, r^2s, r^4s\}.$$

You can easily check that this satisfies the subgroup axioms.

Solution 3.1

We start by considering what can be obtained just from r^2. We must have all the integer powers of r^2, that is

$$r^2, \quad (r^2)^2 = r^4 \quad \text{and} \quad (r^2)^3 = r^6 = e.$$

So far, we have all the elements of the cyclic *subgroup* $\langle r^2 \rangle$, so further combinations will give nothing new. (The closure axiom applied to $\langle r^2 \rangle$ ensures this.)

If we now combine what we have with s (on the right), we get, in addition,

$$r^2s, \quad r^4s \quad \text{and} \quad es = s.$$

You may recognize the set of elements that we now have,

$$\{e, r^2, r^4, s, r^2s, r^4s\},$$

as forming the symmetry group of the hexagon with inscribed triangle (as considered in the last section). As such, it is closed, and so further combinations will give nothing new.

As we have just seen, the set *is* a subgroup of D_6.

If you did not recognize the set, then you may have calculated all possible combinations. However, in all other cases, we can use the relations in the description of D_6 to reduce the combination to one of those already found. For example,
$$\begin{aligned} sr^2 &= (sr)r \\ &= (r^5s)r \\ &= r^5(sr) \\ &= r^5r^5s \\ &= r^{10}s \\ &= r^4s. \end{aligned}$$

Solution 3.2

Repeatedly combining $t[\mathbf{a}]$ with itself will give rise to all *positive* integer powers of $t[\mathbf{a}]$; that is, we obtain the set

$$\{(t[\mathbf{a}])^n : n \in \mathbb{Z},\ n > 0\}.$$

The set is not a subgroup; although closed, it does not contain the identity element of E_1 or the inverse elements $(t[\mathbf{a}])^{-n}$, $n \in \mathbb{Z}$, $n > 0$.

What are 'missing' from the set are the zeroth and negative powers of $t[\mathbf{a}]$.

Solution 3.3

Some of the arguments in this solution are common to all values of m, some depend on the value of m.

Taking integer powers of r^m will give all the elements of one of the cyclic subgroups $\langle r^m \rangle$ that we have already listed.

When we combine the resulting elements of the cyclic subgroups $\langle r^m \rangle$ with s *on the right*, we get one of the following list of sets (listed according to the value of m):

$m = 1$ $\quad \{e, r, r^2, r^3, r^4, r^5, s, rs, r^2s, r^3s, r^4s, r^5s\} = D_6$;

$m = 2$ $\quad \{e, r^2, r^4, s, r^2s, r^4s\}$;

$m = 3$ $\quad \{e, r^3, s, r^3s\}$;

$m = 4$ $\quad$ as for $m = 2$;

$m = 5$ $\quad$ as for $m = 1$.

When we combine the resulting elements of the cyclic subgroups $\langle r^m \rangle$ with s *on the left*, we can use the relation $sr = r^5 s$ to show that we again get one of the three sets listed above. For example,

$$\begin{aligned} sr^2 &= srr \\ &= r^5 sr \\ &= r^5 r^5 s \\ &= r^{10} s \\ &= r^4 s \quad (\text{since } r^6 = e) \end{aligned}$$

and

$$\begin{aligned} sr^4 &= r^{20} s \quad (\text{using } sr = r^5 s \text{ four times}) \\ &= r^2 s, \end{aligned}$$

and so $\langle r^2 \rangle$ combined with s on the left gives the second set listed above.

Thus there are only three potential subgroups $\langle r^m, s \rangle$ for $m = 1, \ldots, 5$. Of the three sets we have found, we know that the first (D_6) and the second, which we have seen is the same as D_3 (or S_3), are subgroups of D_6. It remains to determine whether the third set is a subgroup.

We know from earlier exercises that each element of this set is its own inverse. Also, the identity element is present. So all that is left to check is the closure axiom. The only combinations of pairs of elements we need check are sr^3, r^3r^3s, r^3sr^3, sr^3s and r^3ss. We have:

$$sr^3 = r^{15}s \quad \text{(using } sr = r^5s \text{ three times)}$$
$$= r^3s;$$
$$r^3r^3s = r^6s$$
$$= s;$$
$$r^3sr^3 = r^6s \quad \text{(since } sr^3 = r^3s\text{)}$$
$$= s;$$
$$sr^3s = r^3s^2 \quad \text{(since } sr^3 = r^3s\text{)}$$
$$= r^3 \quad \text{(since } s^2 = e\text{)};$$
$$r^3ss = r^3s^2$$
$$= r^3.$$

We already know that the other possible combinations of pairs of elements give an element in the set.

Hence the set is closed and is thus a subgroup.

(An alternative approach to showing that this third set is a subgroup is to draw in a diagonal along the reflection axis for s and to note that the elements of the set are precisely the symmetries of the hexagon with this diagonal added.)

If you care to compare these calculations with the calculation of combinations of elements of the Klein group, you will probably believe that we have found yet another representation of V.

Solution 3.4

There are two approaches to this exercise.

The geometric one says that $r^n s$ is a reflection (whatever the value of n) and D_6 is generated by r and *any* reflection.

The algebraic one hinges on the fact that D_6 is generated by r and s. Now

$$\langle r, r^n s \rangle$$

already contains one of these generators, namely r. If we can show that it also contains s, then we have enough to generate all of D_6. So, how could we produce s from r and $r^n s$? Since we have r, we have all integer powers of r, in particular we have r^{-n}, which is the inverse of r^n. Thus we also have

$$r^{-n} r^n s = r^{-n+n} s = r^0 s = s.$$

This completes the proof that $\langle r, r^n s \rangle$ contains both r and s. Hence

$$D_6 = \langle r, s \rangle \subseteq \langle r, r^n s \rangle \subseteq D_6,$$

and so $\langle r, r^n s \rangle = D_6$ for $n = 0, \ldots, 5$.

Solution 3.5

(a) We have found $\langle r^2, s \rangle$ already, in Solution 3.3:
$$\langle r^2, s \rangle = \{e, r^2, r^4, s, r^2s, r^4s\},$$
and is formed by taking the union of $\{e, r^2, r^4\}$ with the results of multiplying each of these on the right by s.

If we replace s by rs in this construction, we obtain
$$\langle r^2, rs \rangle = \{e, r^2, r^4, rs, r^2rs, r^4rs\}$$
$$= \{e, r^2, r^4, rs, r^3s, r^5s\}.$$

Similarly,
$$\langle r^2, r^2s \rangle = \{e, r^2, r^4, r^2s, r^4s, r^6s\}$$
$$= \{e, r^2, r^4, r^2s, r^4s, s\}$$
$$= \langle r^2, s \rangle.$$

As in Solution 3.3, use of the relation $sr = r^5s$ shows that multiplying on the left by r^ns gives us the same sets of elements as does multiplying on the right by r^ns.

In exactly the same way, we find that:
$$\langle r^2, r^3s \rangle = \langle r^2, rs \rangle;$$
$$\langle r^2, r^4s \rangle = \langle r^2, s \rangle;$$
$$\langle r^2, r^5s \rangle = \langle r^2, rs \rangle.$$

Thus, there are just two potential subgroups of the form $\langle r^2, r^ns \rangle$. Both contain the identity element and the inverse of each element. Also both are closed, as we can check in the same way as we checked $\langle r^2, s \rangle$ for closure in Solution 3.3. Thus both are indeed subgroups.

(b) We have found $\langle r^3, s \rangle$ already, in Solution 3.3:
$$\langle r^3, s \rangle = \{e, r^3, s, r^3s\}.$$

Proceeding exactly as in part (a), we find:
$$\langle r^3, rs \rangle = \{e, r^3, rs, r^3rs\} = \{e, r^3, rs, r^4s\};$$
$$\langle r^3, r^2s \rangle = \{e, r^3, r^2s, r^3r^2s\} = \{e, r^3, r^2s, r^5s\};$$
$$\langle r^3, r^3s \rangle = \langle r^3, s \rangle;$$
$$\langle r^3, r^4s \rangle = \langle r^3, rs \rangle;$$
$$\langle r^3, r^5s \rangle = \langle r^3, r^2s \rangle.$$

Thus, there are just three potential subgroups of the form $\langle r^3, r^ns \rangle$. Checking the subgroups axioms, as in part (a), shows that all three are indeed subgroups.

Solution 3.6

The only cases we have not yet considered are when $m = 4$ or 5 (and $n \neq 0$). Since r^4 is the inverse of r^2 and since r^5 is the inverse of r, we must have:
$$\langle r^4, r^ns \rangle = \langle r^2, r^ns \rangle, \quad n = 0, \ldots, 5;$$
$$\langle r^5, r^ns \rangle = \langle r, r^ns \rangle, \quad n = 0, \ldots, 5.$$

Solution 3.7

We can argue geometrically as follows. The two generators given are reflections. The combination of two reflections is a rotation. Thus we must have $(r^m s)(r^n s) = r^k$ (for some $k \in \{0, \ldots, 5\}$). Therefore, $r^m s = r^k (r^n s)^{-1}$, and so an alternative pair of generators for $\langle r^m s, r^n s \rangle$ is r^k and $r^n s$. Hence, by Solution 3.6, $\langle r^m s, r^n s \rangle = \langle r^k, r^n s \rangle$ is one of the subgroups already found.

Alternatively, we can argue algebraically and compute $(r^m s)(r^n s)$ explicitly. We obtain

$$\begin{aligned} r^m s r^n s &= r^m s r r \ldots r s \\ &= r^m r^5 r^5 \ldots r^5 s s \quad \left(\text{using } sr = r^5 s \ n \text{ times}\right) \\ &= r^m r^{5n} e \\ &= r^{m+5n}. \end{aligned}$$

We reduce $m + 5n \bmod 6$, because $r^6 = e$. Therefore the composite $(r^m s)(r^n s)$ is of the form r^k (for some $k \in \{0, \ldots, 5\}$). We can then argue as in the geometric case to deduce that $\langle r^m s, r^n s \rangle$ is one of the subgroups already found.

Solution 4.1

Working modulo 5, we have

$$3^0 = 1, \quad 3^1 = 3, \quad 3^2 = 4, \quad 3^3 = 2, \quad 3^4 = 1.$$

So $\mathbb{Z}_5^* = \langle 3 \rangle$.

Solution 4.2

The product of two typical elements is

$$\begin{bmatrix} 1 & a \\ 0 & 1 \end{bmatrix} \begin{bmatrix} 1 & b \\ 0 & 1 \end{bmatrix} = \begin{bmatrix} 1 & a+b \\ 0 & 1 \end{bmatrix}.$$

This shows that

$$\begin{bmatrix} 1 & 1 \\ 0 & 1 \end{bmatrix}^2 = \begin{bmatrix} 1 & 2 \\ 0 & 1 \end{bmatrix}.$$

Generalizing, we have:

$$\begin{bmatrix} 1 & 1 \\ 0 & 1 \end{bmatrix}^n = \begin{bmatrix} 1 & n \\ 0 & 1 \end{bmatrix},$$

for n a positive integer.

This could be formally proved by the Principle of Mathematical Induction.

Since

$$\begin{bmatrix} 1 & 1 \\ 0 & 1 \end{bmatrix}^{-1} = \begin{bmatrix} 1 & -1 \\ 0 & 1 \end{bmatrix},$$

it follows that

$$\begin{bmatrix} 1 & 1 \\ 0 & 1 \end{bmatrix}^{-n} = \begin{bmatrix} 1 & -n \\ 0 & 1 \end{bmatrix},$$

for n a positive integer.

Thus, generally,

$$\begin{bmatrix} 1 & 1 \\ 0 & 1 \end{bmatrix}^a = \begin{bmatrix} 1 & a \\ 0 & 1 \end{bmatrix},$$

for $a \in \mathbb{Z}$.

This holds even for $a = 0$.

Therefore the group is cyclic with generator

$$\begin{bmatrix} 1 & 1 \\ 0 & 1 \end{bmatrix}.$$

Solution 4.3

(a) $\mathbb{Z}_5$ is closed (by the definition of addition modulo 5) and has identity element 0. The elements 1 and 4 are inverses of each other, as are 2 and 3. So $\mathbb{Z}_5$ is a group.

Since 1 generates $\mathbb{Z}$, it is worth trying 1 as a generator in this case too!

$$0 \cdot 1 = 0$$
$$1 \cdot 1 = 1$$
$$2 \cdot 1 = 1 + 1 = 2$$
$$3 \cdot 1 = 1 + 1 + 1 = 3$$
$$4 \cdot 1 = 1 + 1 + 1 + 1 = 4$$

Hence 1 does generate the whole group and $\mathbb{Z}_5$ is cyclic.

(b) By carrying out similar calculations, we find that each of the non-zero elements 1, 2, 3 and 4 generates the whole group. Thus $\mathbb{Z}_5$ has four different possible choices of generator.

Solution 4.4

We can deduce from our work in Section 3 that no single element of D_6 generates the whole group.

Alternatively we can use the fact that any cyclic group is Abelian. From the description of D_6 in terms of r and s, we know that $rs \neq sr$, so that D_6 is not Abelian and so cannot be cyclic.

Solution 4.5

Because g generates G, every element of G is of the form

$$g^n = \phi(n)$$

for some integer n. Thus ϕ is onto.

Solution 4.6

Suppose that $\phi(m) = \phi(n)$ for some m and n in $\mathbb{Z}$. We must show that $m = n$. Suppose not and that m is larger than n (so that $m - n > 0$). By the definition of ϕ, we have

$$g^m = g^n.$$

It follows that

$$g^{m-n} = e \quad \text{(the identity of } G\text{)}.$$

So, there exist positive powers of g which give the identity. It follows, therefore, that if r is the smallest such, then

$$\langle g \rangle$$

is a finite cyclic group with r elements, contradicting the fact that

$$G = \langle g \rangle$$

is infinite. This contradiction shows that our supposition that $m \neq n$ is untenable, and so $m = n$ and ϕ is one–one.

Solution 4.7

We list the cyclic subgroups below:

$$\langle 0 \rangle = \{0\}$$
$$\langle 1 \rangle = \mathbb{Z}_8$$
$$\langle 2 \rangle = \{0, 2, 4, 6\}$$
$$\langle 3 \rangle = \mathbb{Z}_8$$
$$\langle 4 \rangle = \{0, 4\}$$
$$\langle 5 \rangle = \mathbb{Z}_8$$
$$\langle 6 \rangle = \{0, 6, 4, 2\} = \langle 2 \rangle$$
$$\langle 7 \rangle = \mathbb{Z}_8$$

Solution 4.8

There are quite a number of ways of choosing a pair of elements from 8, but from Solution 4.7 any pair containing 1, 3, 5 or 7 will generate the whole group, which is cyclic.

Taking a pair of elements one of which is 0 is just like taking a single element, and so will generate a cyclic subgroup.

That leaves the case where both elements of the pair are chosen from 2, 4 and 6. We deal with the three possibilities in turn.

Since $\langle 4 \rangle \subseteq \langle 2 \rangle$, we have

$$\langle 2, 4 \rangle = \langle 2 \rangle,$$

and so is cyclic.

A similar observation shows that

$$\langle 4, 6 \rangle = \langle 6 \rangle = \langle 2 \rangle,$$

and so is cyclic.

For the pair 2 and 6, we note that $\langle 2 \rangle = \langle 6 \rangle$, so

$$\langle 2, 6 \rangle = \langle 2 \rangle = \langle 6 \rangle,$$

and so is cyclic.

Solution 5.1

The translation $t[\mathbf{a}]$ moves everything to the right by the width of one tile. Hence $t[\mathbf{a}] \wedge E$ is the edge marked E' in the diagram below.

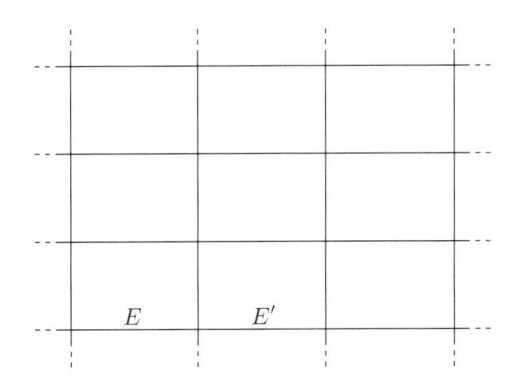

Solution 5.2

Throughout Solutions 5.2–5.4, we use the fact that the elements $r^n s$ of D_6, being reflections, are self-inverse. We also use $sr = r^5 s$ and its generalization

$$sr^k = \left(r^5\right)^k s = r^{5k} s$$

and $r^6 = e$ and its generalization

$$r^{6k} = e.$$

(a) $s \wedge r = srs^{-1} = srs = r^5 s s = r^5$.

(b) $r^n \wedge r = r^n r r^{-n} = r$.

(c) $(r^n s) \wedge r = (r^n s) r (r^n s)^{-1}$
$= r^n s r s^{-1} r^{-n}$
$= r^n s r s r^{-n}$
$= r^n r^5 s s r^{-n}$
$= r^n r^5 r^{-n}$
$= r^5.$

Solution 5.3

(a) $s \wedge s = sss^{-1} = s$.

(b) $r^n \wedge s = r^n s r^{-n}$
$= r^n r^{-5n} s$
$= r^{-4n} s$
$= r^{2n} s \quad$ (since $r^{6n} = e$).

For the various values of n, we get:

$n = 0 \quad r^0 \wedge s = s;$
$n = 1 \quad r^1 \wedge s = r^2 s;$
$n = 2 \quad r^2 \wedge s = r^4 s;$
$n = 3 \quad r^3 \wedge s = r^6 s = s;$
$n = 4 \quad r^4 \wedge s = r^8 s = r^2 s;$
$n = 5 \quad r^5 \wedge s = r^{10} s = r^4 s.$

(c) $(r^n s) \wedge s = (r^n s) s (r^n s)^{-1}$
$= (r^n s) s (r^n s)$
$= r^n s s r^n s$
$= r^n r^n s$
$= r^{2n} s.$

For the various values of n, we get:

$n = 0 \quad r^0 s \wedge s = s;$
$n = 1 \quad r^1 s \wedge s = r^2 s;$
$n = 2 \quad r^2 s \wedge s = r^4 s;$
$n = 3 \quad r^3 s \wedge s = r^6 s = s;$
$n = 4 \quad r^4 s \wedge s = r^8 s = r^2 s;$
$n = 5 \quad r^5 s \wedge s = r^{10} s = r^4 s.$

Solution 5.4

(a) $s \wedge rs = srss^{-1} = sr = r^5 s$.

(b) $r^n \wedge rs = r^n rs r^{-n}$
$= r^{n+1} s r^{-n}$
$= r^{n+1} r^{-5n} s$
$= r^{n+1-5n} s$
$= r^{1-4n} s$
$= r^{2n+1} s.$

For the various values of n, we get:

$n = 0 \quad r^0 \wedge rs = rs;$
$n = 1 \quad r^1 \wedge rs = r^3 s;$
$n = 2 \quad r^2 \wedge rs = r^5 s;$
$n = 3 \quad r^3 \wedge rs = r^7 s = rs;$
$n = 4 \quad r^4 \wedge rs = r^9 s = r^3 s;$
$n = 5 \quad r^5 \wedge rs = r^{11} s = r^5 s.$

(c) $r^n s \wedge rs = (r^n s) rs (r^n s)^{-1}$
$= (r^n s) rs (r^n s)$
$= r^n s r s r^n s$
$= r^n r^5 s s r^n s$
$= r^n r^5 r^n s$
$= r^{2n+5} s.$

For the various values of n, we get:

$n = 0 \quad r^0 s \wedge rs = r^5 s;$
$n = 1 \quad r^1 s \wedge rs = r^7 s = rs;$
$n = 2 \quad r^2 s \wedge rs = r^9 s = r^3 s;$
$n = 3 \quad r^3 s \wedge rs = r^{11} s = r^5 s;$
$n = 4 \quad r^4 s \wedge rs = r^{13} s = rs;$
$n = 5 \quad r^5 s \wedge rs = r^{15} s = r^3 s.$

Solution 5.5

We can deduce the orbits of r, s and rs directly from Solutions 5.2, 5.3 and 5.4 respectively. They are:

$$\operatorname{Orb}(r) = \{r, r^5\}$$
$$\operatorname{Orb}(s) = \{s, r^2 s, r^4 s\}$$
$$\operatorname{Orb}(rs) = \{rs, r^3 s, r^5 s\}$$

Since, for any $g \in D_6$,

$$g \wedge e = g e g^{-1} = e,$$

then $\operatorname{Orb}(e) = \{e\}$.

It remains to consider the elements r^m ($m = 2, \ldots, 5$) and $r^m s$ ($m = 2, \ldots, 5$). As in Solutions 5.2–5.4, we compute as follows:

$$r^n \wedge r^m = r^n r^m r^{-n}$$
$$= r^m$$

$$r^n s \wedge r^m = r^n s r^m r^n s$$
$$= r^n s r^{m+n} s$$
$$= r^n r^{5(m+n)} s s$$
$$= r^{6n+5m}$$
$$= r^{5m}$$

$$r^n \wedge r^m s = r^n r^m s r^{-n}$$
$$= r^{m+n} r^{-5n} s$$
$$= r^{m-4n} s$$
$$= r^{m+2n} s$$

$$r^n s \wedge r^m s = r^n s r^m s r^n s$$
$$= r^n r^{5m} s s r^n s$$
$$= r^{5m+2n} s$$

Hence, substituting in the values $m = 2, \ldots, 5$ and $n = 0, \ldots, 5$ we obtain the following complete list of orbits for D_6 under the given group action:

$$\operatorname{Orb}(e) = \{e\}$$
$$\operatorname{Orb}(r) = \{r, r^5\}$$
$$\operatorname{Orb}(r^2) = \{r^2, r^4\}$$
$$\operatorname{Orb}(r^3) = \{r^3\}$$
$$\operatorname{Orb}(r^4) = \{r^2, r^4\}$$
$$\operatorname{Orb}(r^5) = \{r, r^5\}$$
$$\operatorname{Orb}(s) = \{s, r^2 s, r^4 s\}$$
$$\operatorname{Orb}(rs) = \{rs, r^3 s, r^5 s\}$$
$$\operatorname{Orb}(r^2 s) = \{s, r^2 s, r^4 s\}$$
$$\operatorname{Orb}(r^3 s) = \{rs, r^3 s, r^5 s\}$$
$$\operatorname{Orb}(r^4 s) = \{s, r^2 s, r^4 s\}$$
$$\operatorname{Orb}(r^5 s) = \{rs, r^3 s, r^5 s\}$$

Note that there are only six *distinct* orbits.

Solution 5.6

In the action, tiles map to tiles, edges to edges and vertices to vertices.

Each tile can be mapped to any other by (for example) a suitable translation. Thus the orbit of any tile under the group action is the set of all tiles.

A similar argument can be applied to show that each long edge can be mapped to any other *long* edge (but not to a short edge). Equally, each short edge can be mapped to any other *short* edge. Hence all long edges belong to one orbit, all short edges belong to another.

Finally, each vertex can be mapped to any other by (for example) a translation. Hence all vertices belong to the same orbit.

Thus there are just four orbits: tiles, long edges, short edges and vertices.

Solution 5.7

Closure Suppose that g and h are in $\text{Stab}(x)$. Then

$$\begin{aligned}(gh) \wedge x &= g \wedge (h \wedge x) \quad \text{(by the definition of group action)} \\ &= g \wedge x \quad \text{(since } h \text{ fixes } x\text{)} \\ &= x \quad \text{(since } g \text{ fixes } x\text{).}\end{aligned}$$

Hence gh is in $\text{Stab}(x)$ and $\text{Stab}(x)$ is closed.

Identity One of the consequences of having a group action is that

$$e \wedge x = x$$

for any $x \in X$. This shows that e is in $\text{Stab}(x)$.

Inverses Suppose $g \in \text{Stab}(x)$, then $x = g \wedge x$ and so

$$\begin{aligned}g^{-1} \wedge x &= g^{-1} \wedge (g \wedge x) \\ &= (g^{-1}g) \wedge x \quad \text{(by the definition of group action)} \\ &= e \wedge x \\ &= x.\end{aligned}$$

Hence $g^{-1} \in \text{Stab}(x)$.

This completes the proof that $\text{Stab}(x)$ is a subgroup of G.

Solution 5.8

(a) Since, by the definition of the action,

$$g \wedge e = geg^{-1} = e,$$

the stabilizer of e is the whole of D_6.

(b) Using the results obtained from calculating orbits in Solution 5.2, we have

$$\text{Stab}(r) = \{e, r, r^2, r^3, r^4, r^5\} = \langle r \rangle.$$

(c) Using the results obtained from calculating orbits in Solution 5.3, we have

$$\text{Stab}(s) = \{e, r^3, s, r^3 s\}.$$

OBJECTIVES

After you have studied this unit, you should be able to:

(a) use the group axioms to check whether given sets and associated binary operations define groups;

(b) check whether a given subset of a group is a subgroup;

(c) express products of elements of a group defined by generators and relations in an appropriate standard form;

(d) obtain the symmetry groups of simple geometric figures;

(e) decide whether a given group is cyclic;

(f) identify the subgroup generated by a set of elements of a given group;

(g) check whether a given system is a group action;

(h) identify the orbit of an element of a set X acted on by a group G;

(i) identify the stabilizer of an element of a set X acted on by a group G.

INDEX

Abelian group 24
associativity 6,8
additive group 25
bijection 26
closure 6
conjugate 30
cycle form of permutation 7
cycle notation 7
cyclic group 24–27
cyclic subgroup 17,18,26–27
dihedral group 14,20
general linear group 7
generator 13,18,21–23
group 6
group action 27–32
group axioms 6

homomorphism 28
identity element 6
infinite order of group element 18
inverse element 6
isomorphism 26
Klein group 5
Lagrange's Theorem 23
non-trivial group 17
orbit 30
order of group 23
order of group element 18
partition 31
permutation 7
permutation group 7
stabilizer 31
subgroup 17

subgroup axioms 17
subgroup generated by
  one element 18,21
  two elements 21–23
symmetry group 10
  of plain rectangular frieze
    10–14,19–20
  of plain rectangular tiling 15–16
  of rectangle 8–10
  of regular hexagon 14–15,18,20
trivial group 17
two-line notation 7
$\mathbb{Z}$ 7,25–26
$\mathbb{Z}_n$ 25–27
$\mathbb{Z}_p^*$ 24